H. Mohr

Lectures on Photomorphogenesis

With 219 Figures

Springer-Verlag
Berlin · Heidelberg · New York 1972

Professor Hans Mohr
Biologisches Institut II der Universität
Freiburg i. Br./BRD

For the cover Figs. 2 and 22 of this volume have been used.

ISBN-13: 978-3-540-05879-3   e-ISBN-13: 978-3-642-65418-3
DOI: 10.1007/978-3-642-65418-3

Preface

The discovery of the reversible red far-red control of plant growth and development and the subsequent *in vivo* identification and isolation of the photoreceptor pigment, phytochrome, constitutes one of the great achievements in modern biology.

It was primarily a group of investigators at the Plant Industry Station, Beltsville, Maryland, headed by the botanist H.A. BORTHWICK and the physical chemist S.B. HENDRICKS, who made the basic discoveries and developed a theoretical framework on which the current progress in the field of phytochrome is still largely based.

While the earlier development of the phytochrome concept has been covered by a number of excellent articles by the original investigators [104, 105, 33, 238] as well as by others who joined the field of phytochrome research later [72, 109, 219], a comprehensive and up-to-date treatment of photomorphogenesis is not available at present. Since it seems to be needed for teaching as well as for researchers I have tried to summarize the present state of the field, reviewing the historical aspects of the phytochrome story only insofar as they are required to understand the present situation. The emphasis of my treatment will be on developmental physiology ("photomorphogenesis") rather than on phytochrome *per se*.

The opportunity of writing this review was made possible by a Visiting Professorship granted to me by the University of Massachusetts during the fall term of 1971. The present text is based on a series of 24 lectures which I delivered at U Mass. I am grateful to my colleagues in the Department of Botany at U Mass for their cordial reception and continuous encouragement as well as for the excellent intellectual climate I enjoyed at Amherst.

This book is dedicated to Dr. H.A. BORTHWICK and to Dr. S.B. HENDRICKS. I began to work with the mustard seedling 15 years ago as a foreign postdoctoral fellow at Beltsville. The mode of cooperation among the Beltsville group opened my eyes to the benefits of team-work, and the wisdom, humility and helpfulness of the two senior scientists was an unforgettable experience which has been a constant inspiration ever since.

Freiburg i. Br. July 1972   H. MOHR

Contents

Use of Symbols

Unfortunately, there is no rigorously defined use of symbols and abbreviations in experimental biology. The S.I. (Système International d'Unités), though formally approved in 1960 by the Conférence Générale des Poids et Mesures, is being accepted in practice only slowly and sometimes even reluctantly. As a compromise the "General Rules for Abbreviations" as defined in every January issue of Plant Physiology were followed in the present text in cases where it seemed too premature to follow strictly the S.I. In general dimensions and concentrations are denoted by brackets. In the following list those abbreviations and symbols are put together which are not trivial and which are repeatedly used in the text.

Symbol (or Abbreviation)	Meaning
$A (= O.D.)$	absorbance
$\Delta A (= \Delta O.D.)$	change in absorbance
$\Delta \Delta A (= \Delta (\Delta O.D.))$	change in change of absorbance
Act. D	Actinomycin D
ALA	δ-amino-levulinic acid
AO	ascorbate oxidase (EC 1.10.3.3.)
CAP	chloramphenicol
d	day
DGD	digalactosyl diglyceride
E	Einstein (= mole quanta)
$E \cdot cm^{-2} \cdot s^{-1}$	quantum flux density
fr	far-red light
GA_3	gibberellic acid
GPD	glyceraldehyde-3-phosphate dehydrogenase
h	hour
HIR	high irradiance response
LOG	lipoxygenase (EC 1.99.2.1.)
$lx (= \frac{lm}{m^2})$	illuminance
M	molar (= mole $\cdot$ liter^{-1})
MGD	monogalactosyl diglyceride
min	minute
mole	mole (a gram molecule)
μ	micro . . . (= 10^{-6})
n	nano . . . (= 10^{-9})
nm	nanometer (= 10^{-9} m)

p	pico . . . ($= 10^{-12}$)
P	phytochrome
P_r	red absorbing form of phytochrome
P_{fr}	far-red absorbing form of phytochrome
[P]	brackets in connection with phytochrome denote concentration
P_{fr}^*	some excited species of P_{fr}
$P_{fr\,(ground\ state)}$	P_{fr} in the ground state, that is, non-excited state
$P_{total} = P_{tot}$	total phytochrome, that is, $[P_r] + [P_{fr}]$
PAL	phenylalanine ammonia-lyase (EC 4.3.1.5.)
r	red light
s	second
$\tau\,\frac{1}{2}$	half-life
UV	short wavelength ultraviolet (< 300 nm)
$W \cdot cm^{-2}$	irradiance (light intensity)
X	unknown reactant of phytochrome (P_{fr})

Some plants which are often mentioned in the text

buckwheat	*Fagopyrum esculentum* Mach.
common male fern	*Dryopteris filix mas* (L.) Schott
lettuce	*Lactuca sativa* L.
maize, corn	*Zea mays* L.
(white seeded) mustard	*Sinapis alba* L.
oat	*Avena sativa* L.
pea	*Pisum sativum* L.
rye	*Secale cereale* L.

Phenomenology of Photomorphogenesis; the Goals of Photomorphogenic Research; the Operational Criteria for the Involvement of Phytochrome

All living systems on this planet depend on a narrow band in the electromagnetic spectrum which we call "light". By the term "light" we designate that range of the electromagnetic spectrum which causes the sensation of light in man. Physically this range can be located between about 390 and 760 nm. When we deal with plants we will also use the term "light" in the sense of "visible radiation" but we usually include the near ultraviolet down to about 320 nm and the very near infrared up to about 800 nm.

Quanta of this spectral range have a relatively low energy of about 90 down to 35 Kcal · Einstein^{-1} (5.5 down to 2.5 eV · photon^{-1}), depending on the particular wavelengths. This energy is sufficient to alter the outer electronic energy levels of atoms or molecules but not sufficient to complete ionization. Living systems have adapted themselves to this type of radiation in course of evolution. Quanta of this size can be absorbed in a plant or in an animal only by a very few types of molecule which are characterized by extended π-electron systems, such as for example the chlorophylls or the carotenoids. Most molecules which occur in the cell (water, proteins, nucleic acids, lipids, carbohydrates and their metabolites) cannot absorb quanta in the spectral range of light so as to result in an electronic excitation.

All life on earth is fuelled by sunlight, that is, by photosynthesis performed by plants. In this process light supplies the free energy to make those organic molecules of which all living systems are principally composed. Those plants and animals which are incapable of photosynthesis live at the expense of photosynthetic plants. In this sense, photosynthesis is the most important biological process on earth – but the influence of light on living systems is much more complex than this. Human vision has been mentioned already, and we also find the vision of all other animals, the bending of plants and plant organs toward light, i.e. phototropism; the oriented movements of animals or plants toward or away from light, i.e. phototaxis; and we find "photomorphogenesis".

By this term, "photomorphogenesis", we designate the fact that light can control growth and differentiation (and therewith development) of a plant independently of photosynthesis. In order to grasp the full importance of this phenomenon we have to remember that the specific development of a living system depends on the genetical information of the particular system and on its environment. The most important factor of the environment is light, at least in all higher plants. It is important, however, to realize that even light does not carry any specific information. However, light can be regarded as an "elective" factor which influences the manner in which those genes are used which are present in the particular organism. In this sense the study of photomorphogenesis is part of a world-wide program to investigate the influence of the environment on the development of higher organisms, including man. This topic of "developmental genetics and environment" is not only a scientific matter. Obviously there are great practical and even political implications. We must know, for instance, the physiological laws which govern the interaction of genes and environment in order to be able to improve (and eventually optimize) our educational system. Higher plants are very probably the systems best suited to identify these general laws.

Fig. 1, showing mustard seedlings, illustrates the basic phenomena of photomorphogenesis. All three seedlings have virtually the same genes, the same chronological age (72 h after sowing at 25° C), and all three were grown on the same medium. Light must be responsible for the obvious differences in morphogenesis of the etiolated and the light-

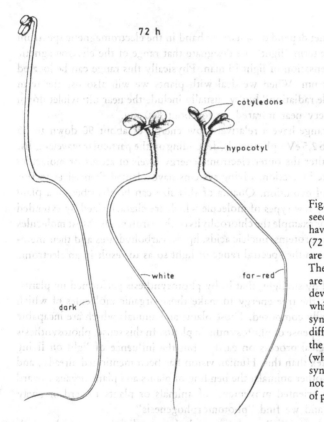

Fig. 1. These seedlings of white-seeded mustard (*Sinapis alba*) have the same chronological age (72 h after sowing at 25 °C) and are virtually identical genetically. The differences in morphogenesis are due to the light. Since the development of the seedling under white light (which allows photosynthesis) does not significantly differ from the development of the seedling under far-red light (which does not support photosynthesis) the effect of light may not be considered a consequence of photosynthesis

grown seedlings. Since the development of the seedling under white light (which allows photosynthesis) does not significantly differ from the development of the seedling under continuous far-red light (which does not support photosynthesis), the effect of light on morphogenesis may not be considered a consequence of photosynthesis. A second example (Fig. 2): You possibly know that the sprouts of a potato tuber will etiolate in darkness whereas in the light the normal potato plant will develop. A characteristic of etiolation is that the internodes grow rapidly while the leaves remain rudimentary. The biological or ecological significance of etiolation is obvious. As long as a plant has to grow in darkness it uses the limited supply of storage material predominantly for axis growth. In this way the probability is highest that the tip of the plant will reach the light before the storage material is exhausted. The physiological problem is to understand the mechanism whereby light can force a plant to develop in a "normal" manner instead of performing etiolation. Apparently two things are involved: firstly a photochemical reaction, and secondly a change

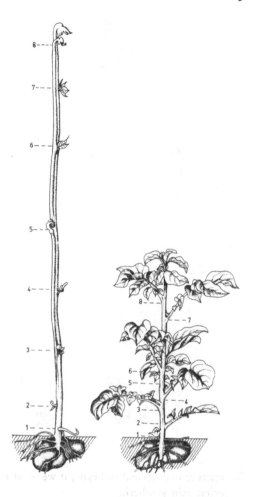

Fig. 2. These two potato plants (*Solanum tuberosum*) are genetically identical. Nevertheless, the dark-grown and the light-grown plants differ greatly (after [200])

in the manner in which the genetic information (i. e. the metabolic potential of the particular organism) is actually used. We recognize that the causal analysis of photomorphogenesis may significantly contribute to our knowledge of development and lead to a better understanding of a basic problem of traditional as well as molecular biology, i.e. differentiation and development of multicellular systems.

Seedlings of dicotyledonous plants (e. g. *Sinapis alba*, mustard, or *Lactuca sativa*, lettuce) turned out to be excellent subjects for the causal analysis of photomorphogenesis. The first stages of vegetative growth after seed germination are especially useful for investigations on photomorphogenesis for the following reasons (Fig. 3): Firstly, the seedling (of *Sinapis alba* in the present case) consists during this phase of only three parts, cotyledons, hypocotyl, and radicle (or taproot). The plumule is hardly developed. Secondly, the seedling contains so much storage material – mainly fat and protein – in the cotyledons that it is completely independent of photosynthesis or of the external supply of organic molecules for a number of days. Thirdly, the seedling can be grown on a medium which supplies only water. An external supply of ions is not required. The seedling can therefore be regarded as a closed system for nitrogen or phosphate. This is a considerable advantage

Fig. 3. Stages of the ontogeny
of a typical dicotyledonous
plant (*Sinapis alba*, white-
seeded mustard). Only sporo-
phytic stages are shown. The
processes of seed germination,
vegetative development and
flower initiation are especially
liable to control by light
(photomorphogenesis)

(for instance over animal embryos) if we want to study light-induced changes of protein
or nucleic acid synthesis.

If we look at the total ontogeny of a typical dicotyledonous plant (represented by
white-seeded mustard, *Sinapis alba*, in the present case) we can identify at least three developmental stages where photomorphogenesis, that is, the control of development by light,
does in fact occur: seed germination, the seedling's development, and the transition from
the vegetative stage to the flowering stage.

Our present problem is to understand the causalities of photomorphogenesis in a modern, preferentially "molecular" terminology. In order to approach the problem we ask
the following questions: What type of pigment absorbs the effective light, what kinds
of photochemical reactions occur, and what are the causal relationships between the photochemical reactions and the final photoresponses which are clearly visible and measurable,
for example, the light-induced growth of the cotyledons, the light-induced synthesis of
pigments like the red anthocyanins or the green chlorophylls, or the light-induced inhibition
of hypocotyl lengthening? And the final question: How can the integration of the different
photoresponses be understood? This integration of the photoresponses of the different
cells, tissues and organs is essential, for photomorphogenesis is a complex but harmonic
process. The integration of the different photoresponses must be precisely regulated in

space and time. Altogether this is a most difficult problem, and the question arises what approach will be feasible at all at present.

We assume that development is primarily the consequence of an orderly sequence of changes in the enzyme complement of an organism. Therefore, the investigator of photomorphogenesis will try to explore those responses in which changes in enzyme levels have a well-defined causal role in well-defined developmental steps.

To return to the first question – what type of pigment absorbs the effective radiation? – the answer is phytochrome (at least, above 500 nm). There is no need to describe the history of phytochrome during the last decade in detail. It is a fascinating story but has already become a classic: the physiological detection, the indirect biophysical characterization and finally the spectrophotometric assay *in vivo* and *in vitro* and the biochemical isolation of phytochrome. This part of the story was closely connected with the existence of the so-called Beltsville group, headed by HARRY BORTHWICK and STERLING HENDRICKS.

The phytochrome story has been described by the original investigators and others on several occasions (cf. Preface). We shall mention only selected results from the broad field of "classical" phytochrome research in order to give some basis for a treatment of the present state.

The phytochrome system is widely distributed in the plant kingdom somewhat like chlorophyll a and it is as basically important for photomorphogenesis as chlorophyll a is for photosynthesis. Essential features of the phytochrome system may be briefly illustrated by using the classical example of induction of germination by light. Lettuce seeds of a light-sensitive variety are sown on a suitable medium at 25 ° C and kept under white fluorescent light. After 24 hours all the seeds have germinated. If the seeds are placed in darkness after sowing, virtually no germination occurs. Most of the seeds are "obligate light germinators". It can be shown that even a short illumination of the soaked seeds, for example, for a few minutes with white fluorescent light of low intensity, can induce germination of the total seed population. Placed in darkness after the light flash, the seeds germinate completely. Having made this observation, namely that germination can be *induced* by light, we can determine the action spectrum of this induction of germination, that is, we can determine the relative efficiency of the different parts of the spectrum applied.

It should perhaps be emphasized that correct action spectra of photoresponses which present the quantum responsivity as a function of wavelength are among the most effective tools of modern biology, ever since WARBURG used this technique to identify cytochrome oxidase. In most cases the action spectrum of a response is the only means of identifying the active photoreceptor in a living system or of finding out if two or more photoresponses are mediated by the same photoreceptor. The approach is indicated in principle in Fig. 4. We must determine dose-response curves at different wavelengths. This task requires relatively large fields of monochromatic radiation of variable quantum flux densities. The proper selection and application of modern physical methods of radiation production, control and measurement requires a detailed knowledge of appropriate physical techniques. In the U.S.A. high-power spectrographs have been developed for biological purposes; on the other hand, we decided in due course to construct our irradiation equipment on the basis of interference filters using high-power xenon arcs as sources of radiation.

On the basis of experimental dose-response curves we can calculate the number of quanta (or Einsteins) per cm^2 (N) which are required at each wavelength to produce the same physiological response (a). The reciprocal values of N as a function of wavelengths are called the "action spectrum" of the particular response we have measured. A relatively

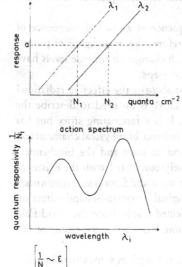

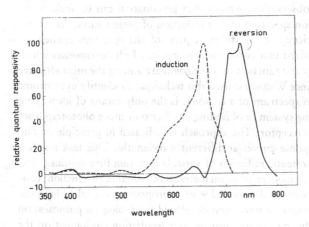

Fig. 4. The elaboration of action spectra proceeds in two steps. First, experimental determination of dose-response curves (above); second, calculation of quantum responsivity as a function of wavelength (below)

simple mathematical treatment shows that an action spectrum represents the extinction or absorption spectrum of the photoreceptor involved if the Law of Reciprocity is valid (i. e. the effect is determined by the product irradiance × time) and if several further prerequisites, which can be checked, are fulfilled. It is obvious that identical action spectra of different photoresponses mean that they are mediated by the same photoreceptor.

If we determine the action spectrum of the photoresponse "germination of lettuce seeds" we find that it is mainly radiation between 550 and 700 nm which induces germination (Fig. 5). Radiation around 660 nm is the most effective. We call this spectral range between

Fig. 5. Typical action spectra of photoresponses due to the phytochrome system. These spectra were determined by WITHROW et al. [298] for the light-dependent opening of the plumular hook of the bean seedling. They are representative of other action spectra of responses *induced* by phytochrome, lettuce seed germination for example

550 and 700 nm "red". Radiation above 700 nm does not induce germination. At first sight it seems that this radiation is without effect, but if we induce 100% germination with red light and then irradiate immediately after the red with a corresponding dose at a wavelength above 700 nm, for instance 730 nm, virtually no germination occurs. In other words, the induction of germination by red light can be nullified by a subsequent irradiation

with the wavelength 730 nm, and is thus reversible. If we determine the action spectrum of the reversing effect, we find radiation between about 700 and 800 nm to be effective (Fig. 5). The highest activity is around 730 nm. We call this spectral range between 700 and 800 nm "far-red". The principle of this reversible red, far-red reaction system is illustrated by Fig. 6. In the dark only a few seeds germinate, but 1 min of red light leads to

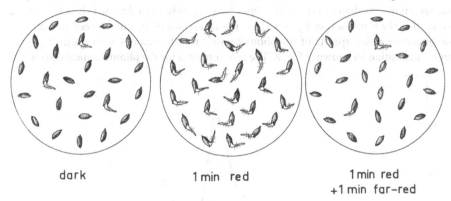

dark   1 min red   1 min red
                                      +1 min far-red

Fig. 6. Reversion of an induction by red light of lettuce seed germination by a subsequent irradiation with far-red light

100% germination which can be detected 24 h later. Following the red with 1 min of far-red represses germination. Following the far-red again with red light would lead to 100% germination, and so on.

It has been shown over the years that this "reversible red, far-red reaction system" not only plays the decisive role in light-induced seed or spore germination but that it is involved in a very great number of photoresponses among plants, from flagellates like *Chlamydomonas*, green algae, mosses, liverworts and ferns up to the di- and monocotyledons. In all cases one can induce a photoresponse – for instance the respiration of *Chlamydomonas eugametos* or the enlargement of the cotyledons of a dark-grown seedling – by red and reverse this induction more or less completely by immediately following with a corresponding dose of far-red light. The common photoreceptor system involved is nowadays called "Phytochrome".

Phytochrome is a bluish chromoprotein having two interconvertible forms, phytochrome 630 or P_r with an absorption maximum in the red at 660 nm and phytochrome 730 or P_{fr} with an absorption maximum in the far-red at 730 nm (Fig. 7). In a dark-grown seedling only P_r, the physiologically inactive species, will be present. The physiologically active P_{fr} can only originate under the influence of light. If the system is irradiated with

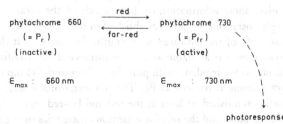

phytochrome 660 $\xrightarrow{\text{red}}$ phytochrome 730
$(=P_r)$ $\xleftarrow{\text{far-red}}$ $(=P_{fr})$
(inactive)   (active)

ε_{max} : 660 nm ε_{max} : 730 nm

Fig. 7. A simple formulation for the phytochrome system   photoresponses

red light a large part of the P_r will be transformed into P_{fr}. This is a complex reaction which has been intensively investigated. The only photochemical event is the excitation of P_r, which leads to a common intermediate. This intermediate gives rise in the dark to others of short life-times all of which finally become P_{fr}. On the other hand an irradiation with far-red which excites P_{fr} can bring this form back to the status of P_r. In this case also, several intermediates of short life-times have recently been detected. It is interesting to note that the formation of P_{fr} as well as the formation of P_r can proceed in solution. The respective action spectra of the photochemical transformations with purified phyto-chrome in solution are shown in Fig. 8. The action spectra for the photochemical conversion

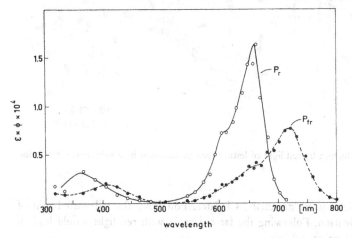

Fig. 8. Action spectra of photochemical transformations of P_r and P_{fr}. The molar extinction coefficient, ε, is in liter $\cdot$ mol^{-1} $\cdot$ cm^{-1} and the quantum yield, Φ, is in mol $\cdot$ Einstein^{-1} (after [32])

of P_r to P_{fr} and P_{fr} to P_r in solution coincide neatly with the action spectra *in vivo*. The action spectrum in solution is given as the product of the molar extinction coefficient, ε, and the quantum efficiency, Φ, as a function of wavelength. Since the phototransformations *in vitro* are known to be first order, the products $\varepsilon \cdot \Phi$ can be determined at any wavelength from the first order rate constants. The value of the first order rate constant was determined experimentally from the slope of a semilogarithmic plot of per cent conversion versus time of irradiation. One quantitative feature of Fig. 8 may be emphasized: The molar extinction coefficient, ε, of P_r at 665 nm is at least 1.6×10^4 litre $\cdot$ mole chromophore$^{-1} \cdot$ cm^{-1}, or greater if the quantum efficiency is less than one.

The absorption spectra of P_r and P_{fr} overlap considerably (Fig. 9). This overlap is the reason why, under conditions of saturating irradiations, photostationary states are characte-ristic for the status of the phytochrome system in solution as well as in the cell. Fig. 10 gives some information on how much of the total phytochrome is present as P_{fr} if the photostationary state is established in the tissue of the mustard seedling by monochro-matic light of the indicated wavelengths. We see that at the most about 80% of the total phytochrome is available as P_{fr}, for instance if we irradiate with pure red light around 660 nm. If we irradiate with pure far-red around 730 nm only about 2 or 3% of the total phytochrome is present as P_{fr}. The photostationary state of the phytochrome system is rapidly established, at least in the red and far-red where the extinction coefficients of the phytochromes and the relative quantum efficiencies of the photochemical transformations

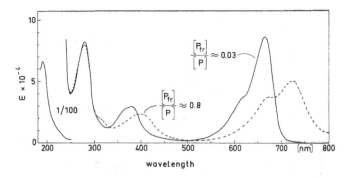

Fig. 9. Absorption spectra of a highly-purified preparation of oat phytochrome after a saturating irradiation with red or far-red light. The photostationary states which will be established by red or far-red light are indicated. The absorbance at wavelengths shorter than 300 nm was determined in a solution diluted 100-fold. Note the asymmetry of the absorption bands which is especially obvious in the case of P_r. This indicates different transitions of the phytochrome (P_r) molecule even in the long wavelengths range (after [180])

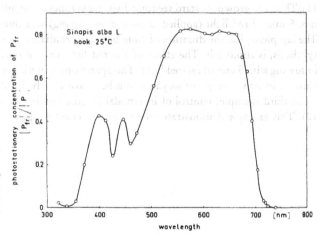

Fig. 10. The percentage of P_{fr} at photoequilibrium as a function of wavelength (*in vivo* measurements at 25 °C with mustard seedlings, hook region) (data from K. M. Hartmann and C. J. P. Spruit [92])

are high. A minute or thereabouts of irradiation with medium quantum flux density is sufficient to establish the photostationary state. In brief, only "short time irradiations" are required to establish photostationary states in the phytochrome system *in vivo* as well as in the photochemically active extract.

On the basis of this information we reach the conclusion that the operational criteria for the involvement of phytochrome in a particular photoresponse can be defined as follows: a photoresponse can be induced by a brief irradiation (e. g. 5 min) with red light of medium quantum flux density; the induction by red light can be fully reversed by immediately following with a corresponding dose of far-red light; the extent of the response following the irradiation sequence red plus far-red must be identical with the extent of the response following a brief treatment with far-red alone.

These operational criteria have been verified in many instances, the classical example being light-mediated seed germination (cf. Fig. 6).

A second example: light-mediated anthocyanin synthesis in the mustard seedling (Fig.

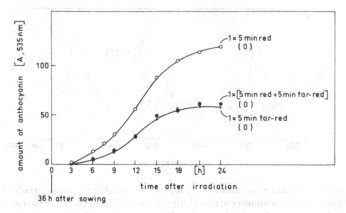

Fig. 11. The time-courses of anthocyanin synthesis in the mustard seedling after brief irradiations with red and/or far-red light (5 min each). Irradiations were performed at time zero, i. e. 36 h after sowing (after [145])

11). The dark-grown mustard seedling does not synthesize significant amounts of anthocyanin. 5 min of red light (applied at 36 h after sowing) will induce anthocyanin synthesis. The lag-phase, i.e. the duration of time between irradiation and the onset of anthocyanin synthesis, is about 3 h. The effect of the red light can be fully reversed by immediately following with 5 min of far-red light. The operational criteria for the involvement of phytochrome in controlling anthocyanin synthesis are clearly fulfilled.

A third example: control of internodal lengthening in normal green bean plants (Fig. 12). This example demonstrates the control exerted by phytochrome over development

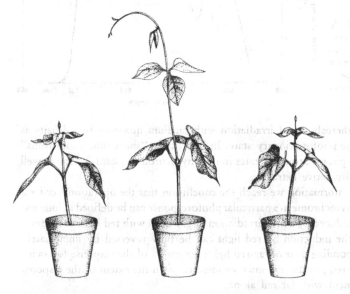

Fig. 12. A demonstration of the controlling function of phytochrome in internodal lengthening in normally-grown green bean plants (*Phaseolus vulgaris*, cv. Pinto). Details of the experiment are described in the text (after [103])

of a normally-grown vegetative plant. We see the changes in internodal lengthening induced by brief irradiations with red or far-red light at the end of an 8 h-day with white fluorescent light of high irradiance (main light period). The Pinto bean on the left received no supplementary irradiation, which means that at the end of every main light period about 80 per cent of the total phytochrome was left as P_{fr} in the plant (the photostationary state established by white fluorescent light does not differ significantly from the photostationary state established by red light). The Pinto bean in the center received 5 min of far-red light at the end of every main light period, and the plant on the right received 5 min of far-red light followed by 5 min of red light at the end of every main light period. The operational criteria for the involvement of phytochrome in control of internodal lengthening are obviously fulfilled.

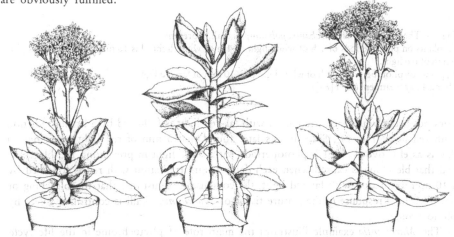

Fig. 13. A demonstration of the controlling function of phytochrome in flower initiation in the short-day plant, *Kalanchoe blossfeldiana*. Details of the experiment are described in the text (after [107])

A fourth example: flowering of the short-day plant, *Kalanchoe blossfeldiana* (Fig. 13). In many short-day plants, a long-day effect can be evoked by the application of a brief illumination in the middle of the dark period in addition to a short-day main light period. The *Kalanchoe* plant on the left was continuously kept under short-day conditions only; it received 8 h of high-irradiance white light per day. The plant in the middle received the same short-day treatment, but in addition received 1 min of red light in the middle of every dark period. The plant remains vegetative. The plant on the right received the short-day treatment plus the 1 min of red light in the middle of the night. However, immediately after the red a 1 min irradiation with far-red light was given. The effect of the red light is fully reversed, the short-day plant flowers. Again, the operational criteria for the involvement of phytochrome are clearly fulfilled: Photoperiodism is controlled through phytochrome.

A fifth example: control of senescence in *Marchantia*. Mature green tissue of the leafy liverwort *Marchantia polymorpha* (gametophytic generation) bleaches markedly when placed in continuous darkness for 4 days but remains green when given daily 1 h photoperiods of white fluorescent light. However, the tissue is "induced" to bleach when each daily 1 h photoperiod is terminated with a brief irradiation with far-red light. The bleaching

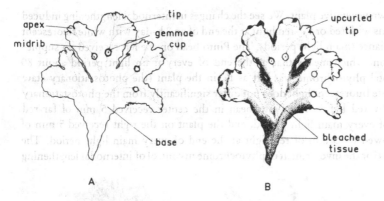

Fig. 14. These drawings of *Marchantia polymorpha* thalli represent
A, plants on photoperiods of 1 h of white light or 1 h of white light plus 10 min of far-red light plus
5 min of red light.
B, plants on photoperiods of 1 h of white light plus 10 min of far-red light
after a 4 day treatment (after [84])

does not occur when each irradiation with far-red light is followed by a brief irradiation
with red light (Fig. 14). Since it was later found that 5 min of red light given once a
day is as effective as the 1 h photoperiods with white light in preventing the bleaching
and that bleaching is caused when each daily 5 min irradiation with red is followed by
a 10 min irradiation with far-red light, the conclusion is justified that the bleaching or
non-bleaching response in the mature tissue of *Marchantia* thalli is controlled solely by
phytochrome.

The *Marchantia* example illustrates the main role of phytochrome in the life-cycle
of a plant: phytochrome is mainly concerned with anabolism, i.e. to permit the build-up
and maintenance of a high metabolic and structural complexity.

While most of the experiments involving phytochrome were done with green algae,
mosses, pteridophytes and spermatophytes, there is experimental evidence that phytochrome is widely distributed in all classes of the plant kingdom, except perhaps in fungi.

In general fungi respond only to blue and ultraviolet light. While some reports about
the effect of long-wavelength visible light on developmental processes in fungi are in print
[146, 28, 35], no analytical work has been done so far. The question of whether or not
phytochrome does control certain stages of fungal development remains open.

Selected Further Reading

HENDRICKS, S.B.: Photochemical aspects of plant photoperiodicity. In: Photophysiology, Vol. 1
    (A. C. GIESE, edit.). New York: Academic Press 1964.
MOHR, H.: Primary effects of light on growth. Ann.Rev. Plant Physiol. *13*, 465 (1962).
SETLOW, R.: Action spectroscopy. Adv. Biol. Med. Phys. *5*, 37 (1957).
WITHROW, R.B. (edit.): Photoperiodism and related phenomena in plants and animals.
    AAAS-Publication No. 55, Washington D.C., 1959.
WITHROW, R.B., WITHROW, A.P.: Generation, control, and measurement of visible and near-visible
    radiant energy. In: Radiation Biology. Vol. 3 (A. HOLLAENDER, edit.). New York: McGraw-Hill
    1956.

Some Properties of Phytochrome

Light controls development and morphogenesis in higher plants through the action of the pigment phytochrome.

Our problem is to understand the nature of phytochrome and the mechanism of phytochrome action at the molecular level. By the term "mechanism" we designate the temporal sequence of elementary steps from light absorption to the final photoresponses. Let us discuss briefly a few characteristics of the phytochrome system proper which are possibly relevant for the discussion of phytochrome action.

From physiological studies, action spectra, kinetics, and reversibility it could be concluded (cf. 1" lecture) that phytochrome must be a bluish substance, having two forms that are interconvertible by light, namely: P_r, with an absorption maximum in the red at about 660 nm, and P_{fr}, with an absorption maximum in the far-red at about 730 nm. P_{fr} is the physiologically active form of the phytochrome system, i.e. the effector molecule (cf. Fig. 28).

On the basis of these predicted spectroscopic properties the Beltsville group (in particular WARREN BUTLER and KARL NORRIS) have developed a method for detecting directly this postulated photoreversible pigment system.

In 1959 the presence of such a pigment was first demonstrated in intact tissue, namely corn coleoptiles, with the aid of a recording, single-beam spectrophotometer designed to measure small optical density changes in a dense light-scattering material. The difference spectrum of the sample (Fig. 15) left no doubt that the reversible red, far-red pigment system really existed. BORTHWICK and HENDRICKS had by that time proposed the name "phytochrome" which was rapidly adopted by all workers in the field.

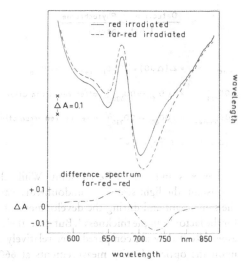

Fig. 15. The first reported difference spectrum (far-red minus red) for phytochrome *in vivo*. The absorbance curves for corn coleoptile tissue were recorded after saturating red or far-red irradiation (after [34])

Since no biochemical assay for phytochrome was known (nor is one so far known) the quantitative assay *in vivo* and *in vitro* was based upon the reversible optical density changes. BUTLER and NORRIS designed a dual wavelength differential photometer which became commercially available under the name "Ratiospect" (Fig. 16). The original ingenious design of the instrument was maintained over the last decade; however, the "Ratiospect" has been greatly improved over the years in sensitivity and reproducibility by physicists in several laboratories. As it is normally used, the Ratiospect measures directly the difference in absorbance (= optical density) of a tissue sample or a solution at two fixed wavelengths

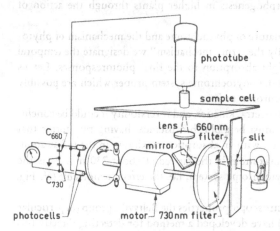

Fig. 16. Diagram of the dual-wave-length difference photometer ("Ratiospect") used for phytochrome assay. Even the original instrument can measure a $\Delta\Delta$ A of 10^{-3} with an apparent optical density of the sample between 2 and 3 (after [15]). More sophisticated double-beam spectro-photometers have recently been built (e. g. by Dr. SPRUIT in Wageningen, Holland). With these instruments it is possible to measure $\Delta\Delta$ A values on the order of $5 \cdot 10^{-5}$

(e.g. 660 and 730 nm (Δ OD $= \Delta$ A)). The total amount of phytochrome can be rapidly estimated from the change in Δ OD following irradiation of the sample with actinic sources of red or far-red light which can drive the photoreaction to completion in the appropriate direction (Fig. 17). $\Delta(\Delta$ OD$) = \Delta\Delta$ A) is a function of the average concentration of phytochrome (c), the sample thickness (x), the light-scattering factor (s), and the extinction coefficient (ε). If $\Delta(\Delta$ OD$)$ is assumed to represent essentially the concentration

Detection of Phytochrome
("Ratiospect")

$P_{total} = \Delta(\Delta OD) = [\Delta OD_{far-red} - \Delta OD_{red}]$

$\Delta OD_{red} = [OD_{660} - OD_{730}]$ after red irradiation

$\Delta OD_{far-red} = [OD_{660} - OD_{730}]$ after far-red irradiation

$\Delta(\Delta OD) = f[c,x,s,\varepsilon]$

Fig. 17. The definitions on which the operation of the dual wavelength differential photometer ("Ratiospect") is based. (OD = A; the term absorbance is recommended to replace optical density)

c, then x, s and ε must be constant. While the extinction coefficient is very probably a constant, the light scattering undoubtedly varies in various kinds of tissues even within the same plant and during the development of a particular tissue. This may also be true for the factor "sample thickness". But nevertheless the Ratiospect data are very trustworthy, especially if they are compared over relatively short periods. To repeat: The assay based upon the optical density measurements at 660 and 730 nm estimates the total amount of phytochrome as the sum of P_r and P_{fr}. Contents of P_r and P_{fr} can be determined separately by measuring the optical density change between 660 and 800 nm or between 730 and 800 nm before and after red or far-red irradiation. At 800 nm neither P_r nor P_{fr} shows any significant absorption (at least by definition). In some cases the 730 vs. 800 nm measurement may be the most reliable assay.

The small amount of protochlorophyllide which is present in etiolated tissue is transformed into chlorophyllide by red irradiation. This causes an increase in the absorbance at 675 – 680 nm (the maximum of the long-wavelength chlorophyll band) while the simultaneous transformation of P_r into P_{fr} results in a decrease in absorbance at 660 nm. The net result is that the decrease of absorbance at 660 nm is less than it would be if only phytochrome responded to the red light. Since the protochlorophyllide → chlorophyllide transformation is irreversible, tissues or preparations in which phytochrome is to be estimated are first irradiated with red light to saturate the protochlorophyllide → chlorophyllide transformation. In studying green plants instead of etiolated tissue the presence of large amounts of chlorophyll makes in vivo measurements of phytochrome extremely difficult and sometimes impossible.

In 1959 photochemically active phytochrome was first isolated by SIEGELMAN, BUTLER and their associates [34] from etiolated maize seedlings by the usual methods of protein extraction in dilute alkaline buffer containing ascorbate and cysteine. In the meantime purification and characterization of phytochrome have been attempted by several laboratories with the following principal results:

1. Phytochrome is a chromoprotein, consisting of a protein moiety and a chromophoric group. According to KROES [140] phytochrome contains no more than one chromophore per protein unit.

2. Phytochrome in solution is stable only between pH 6.5 and 7.5. The reversibility of the pigment is progressively lost as the pH deviates considerably from 7.0. Above pH 8.0 and below pH 6.0 the pigment becomes insoluble and rapidly precipitates [140].

3. The suggestion advanced by the Beltsville group on the basis of physiological data, that the chromophoric group must be an open chain tetrapyrrole, was confirmed by RÜDIGER [225] who used methods of classical organic chemistry (degradation with chromium trioxide) and found that the phytochrome chromophore is indeed a tetrapyrrole very similar to phycocyanobilin, one of the well-known chromophoric groups of the phycobilins of

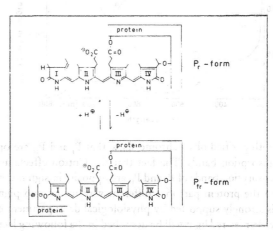

Fig. 18. The most detailed structure proposed so far for the phytochrome chromophore and for the coupling between the chromophore and the protein moiety. The analysis was based on a sample of rye phytochrome (Secale cereale, cv. Balbo) which was denatured (after [225])

red and blue-green algae. However, the chromophore of phytochrome is coupled to its protein by different covalent bonds than is the case with phycocyanobilin. It is very difficult to cleave the phytochrome chromophore from its protein part. A model for the phytochrome chromophore and its binding to the protein moiety is shown in Fig. 18.

Moreover an attempt is being made to characterize the isomerization of the chromophore during the transition from P_r to P_{fr} and *vice versa*. It is proposed that absorption of light by P_r causes a redistribution of charge in the chromophore which is expressed as a shift in the absorption maximum and which results in the outer pyrrole ring with the ethylidine substituent acquiring a proton-donating capacity. This configuration is stabilized by the approach of a proton-accepting group in the peptide chain of the protein, once the conformation of the latter has been changed. The information presently available points to the conclusion that the two reactions, $P_r \rightarrow P_{fr}$ and $P_{fr} \rightarrow P_r$, do *not* have the same intermediates [208, 140]. Nevertheless it is common to the two reactions that the first step is a photochemical change of the chromophore, followed by a conformational change in the protein moiety which no longer depends on light. This is at least true for the visible light. If short-wavelength UV is used (280 nm), the results indicate an energy transfer from the protein to the chromophore [207].

P_{695} is a transient form between P_r and P_{fr} which has been characterized by its absorption peak at 695 nm. P_{695} seems to be an instable isomer of P_r insofar as the photoisomerization of the chromophore has already occurred whereas the conformational changes of the protein moiety have not [140]. Irrespective of these details one may describe the photochemical transformations of phytochrome as a combination of isomerization of the chromophore with a change in the conformation of the protein part. For each of the two chromophore isomers, a protein conformation exists which gives a stable complex of the pigment, P_r in the ground state and P_{fr} in the ground state. Indeed, changes in the optical rotatory dispersion (ORD) (Fig. 19) and circular dichroic spectra (CD) (Fig. 20) were found to correspond with the shifts in the absorption spectra. The ORD and CD measurements

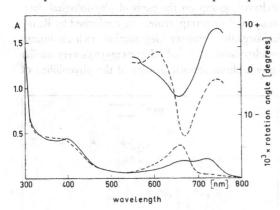

Fig. 19. Absorption spectra and long-wavelength ORD curves of phytochrome. ---, P_r; ——, P_{fr}. Note that the range of the rotation angle is 10^{-2} degrees (after [140])

with purified phytochrome show that P_r and P_{fr} are optically active in their red and blue absorption bands. The fact that the Cotton effects associated with the long-wavelength absorption bands of P_r and P_{fr} are opposite in sign means that the chromophore is attached to the protein part in a different way in the two pigment forms [140]. This conclusion is strongly supported by physiological data; pertinent experiments with linearly polarized red or far-red light will be described in the following lecture. Since no significant differences

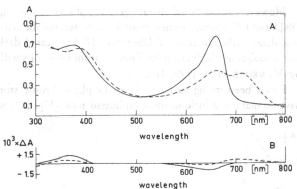

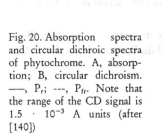

Fig. 20. Absorption spectra and circular dichroic spectra of phytochrome. A, absorption; B, circular dichroism. —, P_r; ---, P_{fr}. Note that the range of the CD signal is $1.5 \cdot 10^{-3}$ A units (after [140])

of the ORD and CD spectra of P_r and P_{fr} in the ultraviolet region ($\lambda < 300$ nm) could be detected [140], the overall change in the conformation of the protein part during the phototransformations of phytochrome must be very small. Indeed, it seems probable that the optical activity of phytochrome is almost exclusively due to asymmetric bonding between the protein moiety and the chromophore which itself has little or no optical activity [140].

The manner of coupling of the tetrapyrrole chromophore to the protein moiety (Fig. 18) is not yet definitely settled. Recently RÜDIGER analyzed *active* phytochrome obtained from MUMFORD and JENNER (cf. Fig. 9). According to KROES [140] he found that ring I was not detached from the protein during oxidation with chromium trioxide and that there was no difference between P_r and P_{fr} in this respect. In both cases, ring IV was converted into methylvinylmaleimide. RÜDIGER concluded that the phytochrome chromophore is coupled to its protein part by means of one acid substituent and by an unknown bond to ring I. The conflicting results of RÜDIGER's first analysis (Fig. 18) must be ascribed to the fact that the first sample was *denatured*.

The investigation of the protein moiety of phytochrome turned out to be a difficult matter and the results obtained by different groups are still controversial.

The most serious problem has been that phytochrome readily loses activity during handling of large volumes of extract. On the other hand, large volumes are required since the original concentration of phytochrome, even in etiolated tissue, is very low.

For some time several laboratories agreed that the molecular weight of the "phytochrome protomer" is close to 60.000 daltons. However, BRIGGS and his associates [77] have recently obtained evidence that this material is in fact a proteolytic product of a substantially larger native molecule. The most recent estimate of the molecular weight of the large species of phytochrome from rye and oat seedlings has yielded a value of about 120.000. However, in very pure extracts of rye a phytochrome form with an even higher molecular weight has been detected [77]. This might be the 14 S species observed previously by CORRELL et al. [39]. The researchers in this field feel that differences between different sources, e.g. between different seed lots, reflect differences in level of contaminant proteolytic enzymes rather than inherent differences in phytochrome. It is noteworthy that even the small molecular weight material is spectrally normal and completely photoreversible. Fig. 21 shows the absorption spectra of a phytochrome extract of low molecular weight after adsorption on calcium phosphate gel. The corresponding difference spectrum is obviously normal.

However, any detailed characterization of 60.000 molecular weight phytochrome (e.g. ORD and CD characteristics) must be repeated on the native molecule. To support this postulate, studies in BRIGGS' laboratory [77] indicated dark reversion properties for the native molecule which are quite different from those of small oat phytochrome as reported by MUMFORD et al. [179, 180, 181].

I have been trying so far to keep the phytochrome story as simple as possible. We have to make it a little more complicated now (Fig. 22): while P_r is stable in darkness,

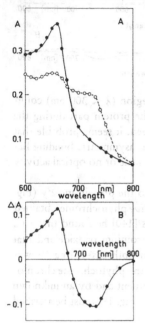

Fig. 21. A, absorption spectra of phytochrome extract after adsorption on calcium phosphate gel. ●——●, after irradiation with far-red light; ○——○, after irradiation with red light. B, corresponding difference spectrum (Δ OD = Δ A) (after [140])

P_{fr}, the physiologically active species of phytochrome, is not. It can disappear in two ways: it is readily destroyed *in vivo* by an irreversible reaction (operationally, decrease of the signal $\Delta\Delta A$), or it reverts to P_r in a thermal reaction. In a seedling the irreversible destruction of P_{fr} seems always to be the dominant process. In monocotyledonous tissue dark reversion

Model of Phytochrome

$$P_v \xrightarrow{{}^{0}k_s} P_r \underset{{}^{1}k_2}{\overset{{}^{1}k_1}{\rightleftharpoons}} P_{fr} \xrightarrow{{}^{1}k_d} P_{fr}'$$

$${}^{1}k_r$$

Photo Steady State
($\dot{P}_r = \dot{P}_{fr}$)

$$\dot{P}_{i} = -k_{1} P_{i} + k_{2} P_{ii} \qquad\qquad \dot{P}_{ii} = +k_{1} P_{i} - k_{2} P_{ii}$$

$$\gamma = \frac{[P_{fr}]}{[P_r] + [P_{fr}]} = \frac{k_1}{k_1 + k_2} = \frac{N_\lambda \cdot \epsilon_r \cdot \phi_r}{N_\lambda \cdot \epsilon_r \cdot \phi_r + N_\lambda \cdot \epsilon_{fr} \cdot \phi_{fr}} = \frac{\epsilon_r \cdot \phi_r}{\epsilon_r \cdot \phi_r + \epsilon_{fr} \cdot \phi_{fr}}$$

Fig. 22. A model of the phytochrome system which includes dark reversion ($P_{fr} \dashrightarrow P_r$) and decay ($P_{fr} \longrightarrow P_{fr}'$) (after [249, 250])

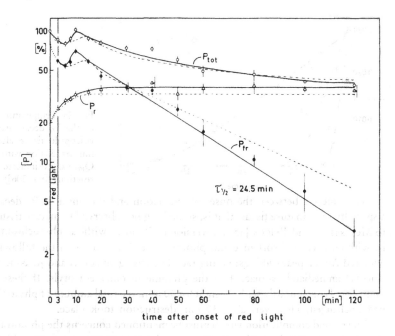

Fig. 23. Phytochrome time-course of 72 h dark-grown mustard cotyledons, following 3 min of saturating red light. ●, measured values of P_{fr}; ○, measured values of total phytochrome (assumption: $[P_{fr}]/[P_{tot}] = 0.8$ for red light). Δ, calculated values of $P_r = P_{tot} - P_{fr}$. The broken lines represent the kinetics for 48 h dark-grown cotyledons (after [150])

has so far never been detected. In the cotyledons of the mustard seedling the decay of P_{fr} can be described (after a lag phase of 10 min) by a first order reaction (Fig. 23). The half life of this first order decay reaction is a function of the age of the seedling. 36 h after sowing the half-life is about 45 min, 48 h after sowing the half-life is about 35 min, 72 h after sowing the half-life is about 25 min (at 25° C) [150]. In dicotyledons, at least, the rate of P_{total} decay is a function of the amount of P_{fr} available for the decay process; under continuous irradiation the decay rate is proportional to the $[P_{fr}]/[P_{total}]$ ratio maintained (Fig. 24). The decay as such is very probably an enzymatic process (Fig. 25).

Fig. 24. Relationship between decay constants for total phytochrome (P_{tot}) and proportion of phytochrome in the P_{fr} form in seedlings of *Amaranthus caudatus* at 25 °C. The meaning of kz: the decay of total phytochrome can be written as: $\log_e P_t/P_0 = - kzt$. When $\log_e P_t/P_0$ is plotted against time the gradient of the line is kz, the rate constant for the decay of total phytochrome (after [132])

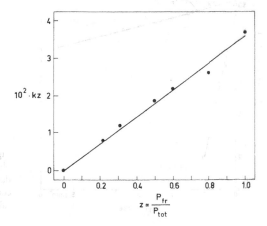

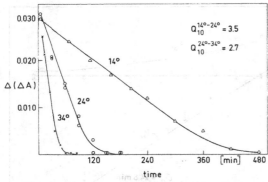

Fig. 25. Temperature dependency of the P_{fr} decay process in *Avena* coleoptile tissue after an initial 5 min irradiation with red light. The Q_{10} values indicate an enzymatic reaction (after [206])

The time lag between the onset of irradiation and the onset of P_{fr} decay is possibly a specialty of immature tissue, that is, something not observed in mature tissue. According to McArthur and Briggs [154], reversion of P_{fr} to P_r without phytochrome decay (that is, without loss of absorbance and photoreversibility) occurs in the following tissues of etiolated Alaska pea seedlings: young radicles, young epicotyls and juvenile region of the epicotyl immediately subjacent to the plumule in older epicotyls. If these tissues were illuminated continuously with red light for 30 min, the total amount of phytochrome remained unchanged. However, after 30 min destruction took place.

The second complication which must be mentioned concerns the phototransformations proper. Schäfer and associates [249] determined dose-response curves (kinetics) for the phototransformation of phytochrome in both directions in tissue from pumpkin and mustard seedlings. The phototransformation kinetics as measured in an optically dense sample deviate from a single first order curve under all circumstances. The kinetics observed have been explained by assuming two populations of phytochrome molecules which are converted with two different rate constants (Fig. 26). Since this result was obtained with different instruments it has been taken as evidence that at least two populations of phytochrome molecules exist *in vivo*, which can be distinguished on the basis of their conversion rates under actinic irradiation. However, this problem is not settled yet either. Spruit and Kendrick [241] have criticized the interpretation of the observed kinetics in terms of two

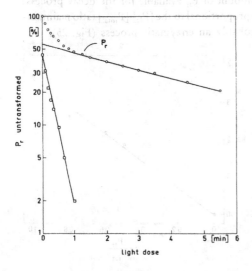

Fig. 26. *In vivo* kinetics of the phytochrome phototransformation ($P_r \longrightarrow P_{fr}$) with 48 h dark-grown mustard cotyledons. ○ , time-course of total P_r; □ , time-course of the calculated fast P_r population (after [249])

or more photochemically different phytochrome populations. They have developed a mathematical model, assuming that the rate of phytochrome phototransformation is proportional to the light intensity and that the light intensity gradient in the sample is exponential. Kinetic curves computed with this model conform closely to measurements. According to SPRUIT and KENDRICK the simplest explanation of the observed kinetics is that there is only one type of phytochrome (as far as the phototransformations are concerned) and that the light intensity gradient in samples that are not too thin is close to exponential.

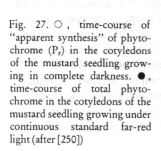

Fig. 27. ○, time-course of "apparent synthesis" of phytochrome (P_r) in the cotyledons of the mustard seedling growing in complete darkness. ●, time-course of total phytochrome in the cotyledons of the mustard seedling growing under continuous standard far-red light (after [250])

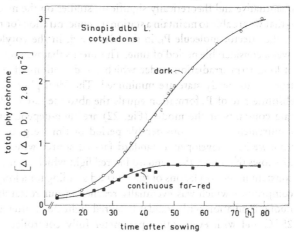

The third complication is that "apparent synthesis" of P_r (operationally, increase of the $\Delta\Delta$ A signal) can continuously occur in many organs, e.g. in the cotyledons of the mustard seedling growing in the dark (Fig. 27). There are at least two ways in which photoreversible phytochrome may appear in the tissues of a germinating seed or a developing seed. 1. Phytochrome is initially present and becomes photochemically functional upon hydration. 2. The pigment (or at least some part of it) is synthesized *de novo*. Increases of the signal $\Delta\Delta$ A which indicate *de novo* synthesis have been observed in several instances, e.g. in seedlings of *Amaranthus caudatus*[133] and in the cotyledons of the mustard seedling (Fig. 27). While growth of the cotyledons (as measured by organ expansion) is almost negligible in continuous dark [175], the apparent increase of phytochrome (P_r) is very strong and follows a sigmoid time-course. There can be little doubt that the increase of the signal $\Delta\Delta$A in the dark-grown mustard cotyledons is very probably exclusively due to the increase in the amount of P_r. Since the rate of increase is nearly constant over a considerable period of time we describe this reaction as a zero order reaction.

Using D_2O for density labelling of the protein moiety of phytochrome, P. QUAIL [210] has shown recently that the increase of the signal $\Delta\Delta$A in the apical hook and in the cotyledons of pumpkin seedlings must be attributed to *de novo* synthesis of phytochrome protein, in the dark-grown seedling as well as in a seedling treated with light and later on placed in the dark ("recovery increase" of $\Delta\Delta$A).

The simplest model of the phytochrome system as it occurs in the cotyledons of the mustard seedling requires four elements and five rate constants even if we ignore the possibility that the phototransformation kinetics reveal more than one P_r and P_{fr} population (Fig. 22).

The model indicates that P_r is formed from a precursor, designated as P_v, through a zero order reaction. The phototransformations follow first order reactions in both directions. The rapid but quantitatively insignificant dark reversion of P_{fr} to P_r follows first order kinetics. The same is true for the irreversible destruction of P_{fr}. To increase the complexity, it has been found that the rate constants may vary in time [250]. This indicates a typical biological system: several elements, several rate constants, and the rate constants changing with time.

Given this background, we must solve the following problem: A prerequisite for most quantitative and theoretically significant studies on the mechanism of phytochrome action is that we are able to maintain a stationary concentration (or nearly stationary concentration) of the effector molecule P_{fr} in the tissue, e.g. in the cotyledons of the mustard seedling, over a considerable period of time. This means that we must experimentally find conditions of long-term irradiation under which the deviations of the phytochrome system from a true photo-steady state are minimized. The ideal photo-steady state is established if the absolute rate of P_r formation equals the absolute rate of P_{fr} disappearance. Since the five rate constants of the model (Fig. 22) are time-dependent, it is difficult to match these requirements over a considerable period of time, e.g. over 24 or 48 h. In the course of time we have developed a standard far-red source which satisfies our requirements. Fig. 27 shows that under the standard far-red light which we use, total phytochrome is virtually constant in the cotyledons of the mustard seedling for a long period of time. The pragmatic compromise which was eventually reached in our research group has been to start continuous irradiation with standard far-red light 36 h after sowing of the mustard seed at 25°C and with every other parameter fully controlled.

Under these conditions the situation of the phytochrome system under continuous standard far-red light can approximately be described as a steady state (Fig. 28). Synthesis of P_r and irreversible decay of P_{fr} compensate to yield a steady state with only a few per cent of the total phytochrome present as P_{fr}. The photo-steady state is virtually established within the order of a minute after the onset of far-red light.

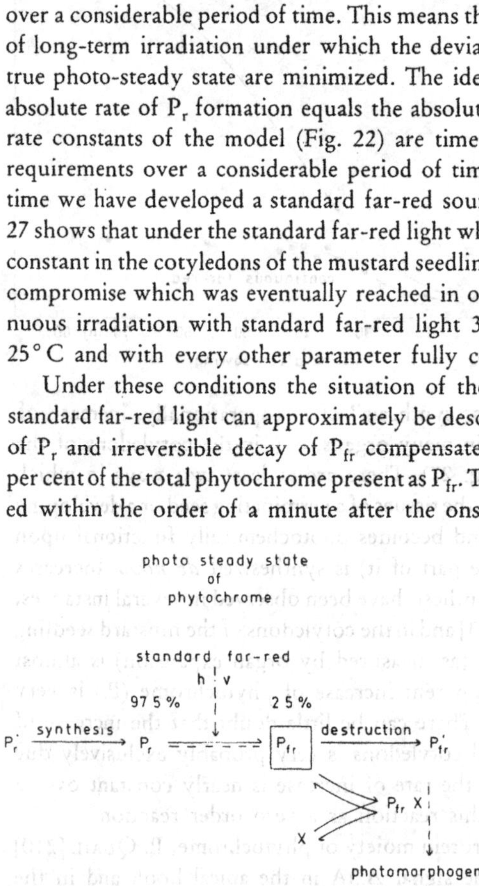

Fig. 28. A model to illustrate the photo-steady state of the phytochrome system which can approximately be maintained in some tissue of the mustard seedling over a considerable period of time by continuous standard far-red light (after [193])

Some further advantages of the standard far-red light are the following: 1) Under these light conditions the formation of chlorophyll a in the mustard seedling occurs very slowly and only in traces (cf. Fig. 60). Since the absorption probability of chlorophyll a in the far-red range is very low moreover, photosynthesis can be ignored even in long-term experiments. 2) Total absorption of far-red light in the cotyledons of the mustard seedling is very low. However, scattering is very strong. These facts lead to an "opal glass effect", i.e. every cell in the cotyledons will receive a similar irradiance, irrespective of the direction and distribution of the incoming far-red light.

These were some selected considerations of the properties of phytochrome as such. The next lecture will be devoted to the question of where in the cell the phytochrome molecules are located.

Selected Further Reading

FURUYA, M.: Biochemistry and physiology of phytochrome. Progr. Phytochem. *1*, 347. London: Interscience Publishers 1968.

KROES, H.H.: A Study of Phytochrome, its Isolation, Structure and Photochemical Transformations. Wageningen: H. Veenman and Zonen N.V. 1970.

ROLLIN, P.: Phytochrome, Photomorphogénèse et Photopériodisme. Paris: Masson, 1970.

SIEGELMAN, H.W.: Phytochrome. In: The Physiology of Plant Growth and Development (M.B. WILKINS, edit.). London: McGraw-Hill 1969.

3ʳᵈ Lecture

Intracellular Localization of Phytochrome

1. Some background information. My first remark is concerned with the interaction of light and molecules [60]. If electronic oscillations are equally possible in all directions, an incident unpolarized light beam will emerge essentially as it entered the system – except for a phase shift. If, however, the particles or molecules in the system are not isotropic and at the same time ordered, an incident unpolarized light beam will be split, because waves oscillating in one direction will have their phases shifted more than those oscillating in another direction. The net result is that anisotropic molecules which are in an order produce two plane polarized emergent beams. Since these emerge in somewhat different directions, the phenomenon is called double refraction or birefringence. The chief use of this phenomenon in biology is in the inverse fashion (Fig. 29): if we line up the molecules of an anisotropic substance then the refraction of plane polarized light measured along and at right angles to the direction of alignment will tell us even more about the electronic structure of the individual molecule.

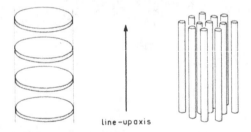

line-upaxis

Fig. 29. Disc-shaped and rod-shaped molecules lined up so that their birefringence may be studied (after [60])

In studies on birefringence the substances in general do not absorb the light. However, similar techniques can be applied to substances which absorb parts of the incident plane (or linearly) polarized light. The results of the analysis are of a similar type as for birefringence. For example, DNA has been studied by irradiating with plane polarized light of wavelength 260 nm, the wavelength of maximum absorption by nucleic acids. There is twice as much absorption perpendicular to the fiber axis as there is parallel to it. Thus one concludes that the absorbing elements themselves lie chiefly perpendicular to the fiber axis. The fact that a structure will absorb plane polarized light differently in different directions is called dichroism. In the following the phenomenon of dichroism is used in order to study the intracellular localization of phytochrome molecules.

Where is phytochrome located inside the cell? – In spite of much effort the techniques of differential centrifugation of the homogenates have so far not yielded a satisfactory answer. Most experiments have led to the findings that phytochrome is easily soluble in watery systems between about pH 6 and 8 and will in general, i.e. in slightly alkaline media, appear in the supernatant. At the most a few per cent of the total phytochrome have been found in the past to be bound to particles or structures. While some claims have recently been made for the localizing of phytochrome in membranes, e.g. in the nuclear envelope [74], the only convincing data on phytochrome localization are based on experiments with linearly polarized light, in other words, are based on studies of dichroism. It was in fact possible to show with the help of linearly polarized red and far-red light in physiological experiments with several types of cells that molecules of phytochrome are oriented in a dichroic structure in the outer cytoplasm, probably along the plasmalemma, the outer cytoplasmic boundary. The phenomena of polarotropism can only be understood if we assume that phytochrome molecules can be highly oriented in the cell.

2. *Polarotropism of fern germlings.* Fig. 30 illustrates in the case of filamentous fern sporelings the basic phenomena of polarotropism. The basic observation – originally made by H. Etzold [61] – is that linearly polarized red light applied from above can strictly determine the direction of growth of a filamentous sporeling growing under special condi-

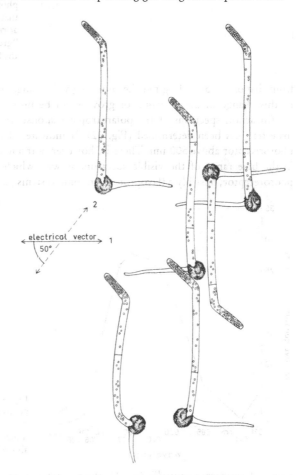

Fig. 30. A basic phenomenon of polarotropism. Linearly polarized light applied from above can strictly determine the direction of growth of a fern protonema (filamentous fern sporeling). The sporelings of *Dryopteris filix-mas* grow normally to the plane of vibration of the electrical vector (redrawn after photographs taken by H. Etzold)

tions on an agar surface. The sporelings will grow at an angle of 90°, that is, "normal" to the plane of vibration of the electrical vector of the linearly polarized light. If we change the plane of vibration the sporelings will rapidly and correspondingly change their direction of growth. We know from other experiments that growth of this filamentous system is restricted to the upper few μm of the filament, that is, to the extreme tip of the apical cell. We know also that the change of the direction of growth is due to the shift of the growing point to the flank of the apex. An accelerated growth on the convex side of the apex is *not* involved (Fig. 31). In order to determine an action spectrum of the polarotropic response we use in principle the following experimental procedure: the sporelings which have been grown so far under polarized red light are irradiated with monochromatic polariz- ed light instead of red. If the plane of vibration of this monochromatic light is different

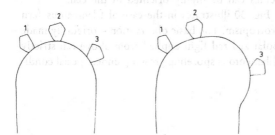

Fig. 31. Starch grains of rice are placed at the tip of the fern protonema to mark certain positions. Left, before and right, after the onset of the polarotropic (or phototropic) curvature. Is is obvious that the polarotropic curvature must be attributed to a shift of the actual "growing point" from the very tip to the flank of the apex (after [61])

from that of the preceding red the sporeling will change its direction of growth. The rate of this change in the direction of growth can be measured with high accuracy.

An action spectrum of the polarotropic response of the sporelings of the common male fern has been determined (Fig. 32). It indicates that phytochrome is the effective photoreceptor above 500 nm. There is, however, a tremendous polarotropic effectiveness in the blue range of the visible spectrum as well which must be attributed to another photoreceptor, possibly a flavoprotein. In some systems (e. g. in the sporelings of the liver-

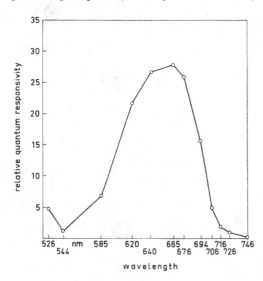

Fig. 32. The action spectrum of the polarotropic response of the fern spore- lings (*Dryopteris filix-mas*) as deter- mined for a response curvature of 19 ° above 500 nm (after [61]).

wort *Sphaerocarpus donnellii*, Fig. 33) the polarotropic response is mediated only through this blue-light-dependent photoreceptor (Fig. 34). We shall return to this phenomenon in a separate lecture. At the moment, the polarotropic response is used only as a tool to investigate the intracellular location of phytochrome, and for this reason our consideration is restricted to those instances where the polarotropic bending is mediated by phytochrome. Fig. 35 represents the situation at the tip of a fern sporeling with respect to phytochrome.

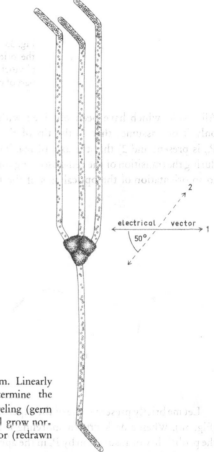

Fig. 33. Another basic phenomenon of polarotropism. Linearly polarized light applied from above can strictly determine the direction of growth of a filamentous liverwort sporeling (germ tubes of *Sphaerocarpus donnellii*). The sporelings will grow normally to the plane of vibration of the electrical vector (redrawn after photographs taken by A. STEINER)

Fig. 34. Action spectra for polarotropism in germlings of a fern (*Dryopteris filix-mas*) and a liverwort (*Sphaerocarpus donnellii*). The action spectrum for the protonema of *Dryopteris* was calculated from dose-response curves at a constant irradiation time of 240 min for the response angle of 17.5°. The action spectrum for the germ tubes of *Sphaerocarpus* was calculated from dose-response curves obtained at a constant irradiation time of 180 min for the response angle of 22.5°. Both angles represent 50 per cent maximum response (after [244])

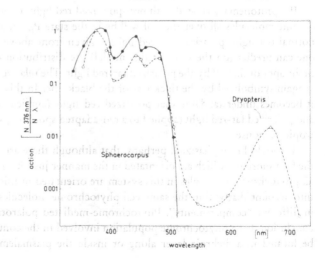

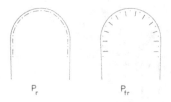

Fig. 35. This model (tip of the fern protonema) may illustrate the orientation of the axis of maximum absorption of the phytochrome chromophore. Left, only P_r present; right, part of phytochrome present as P_{fr} (after [61])

All results which have been obtained with this system can be satisfactorily interpreted only if one assumes that 1) the tip of the apical cell will grow at the site where most P_{fr} is present and 2) that the axis of maximum absorption of phytochrome turns by $90°$ during the transition of the red-absorbing form P_r to the far-red-absorbing form P_{fr}, leading to an orientation of the optical axis of the P_{fr} molecules perpendicular to the cell surface.

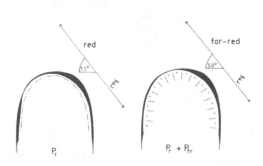

Fig. 36. In this model (tip of the fern protonema) the thickness of the outer line represents the relative amount of P_{fr} present at the different parts of the tip of the apical cell following an irradiation (from above) with polarized red or far-red light. The red light was applied to a dark-adapted protonema (P_r only), whereas the far-red light was applied to a protonema which had previously received normal red light ($P_{fr}/P_{tot} \approx 0.8$). In both cases the same distribution of P_{fr} in the apex will result and thus the same response will appear (after [61])

Let me briefly present some of the experimental results which support this interpretation (Fig. 36). When a dark-grown sporeling is irradiated with polarized red light from above the probability of absorption by P_r in the apex can be predicted as indicated by the thickness of the black line. Consequently the apex will grow normally to the plane of vibration of the electrical vector of the polarized red light.

If a protonema is treated with non-polarized red light a high percentage of the phytochrome molecules all over the cell will be in the state P_{fr}. If, after this pretreatment with normal red light, polarized far-red light is given from above as indicated in the figure, one can predict (on the basis of this model) the distribution of P_{fr} which is determined in the apex of the cell by the polarized far-red light. The relative density of the P_{fr} molecules is again symbolized by the thickness of the black line. In this way (and only in this way) it becomes understandable that polarized red light (applied to a dark-grown sporeling) and polarized far-red light (applied to a red-adapted sporeling) will lead to the same polarotropic response.

It should be emphasized, perhaps, that although there are phytochrome molecules in the fern sporeling which are orientated in the manner just described, it is doubtful whether all phytochrome molecules in this system are orientated in this way. We must rather take into account that within the same cell phytochrome molecules will be located differently in different "compartments". Phytochrome-mediated polarotropism will only allow the conclusion that a phytochrome population involved in the control of cellular growth must be located in a dichroic layer along or inside the plasmalemma.

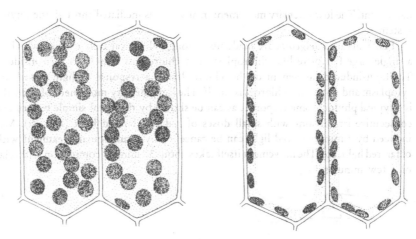

Fig. 37. Diastrophe (left) and parastrophe (right) orientation of the chloroplasts in the cells of a moss (*Funaria hygrometrica*). The cells were irradiated from above with weak (left) or strong (right) light

3. *Chloroplast movement in Mougeotia cells.* Chloroplasts commonly undergo two types of movement in response to light (Fig. 37). The first is called the low intensity movement. It leads to the diastrophe orientation of the chloroplasts and allows the chloroplast to absorb maximum light under normal illumination. The second, or high intensity movement, protects the chloroplast from damage in direct sunlight. This movement leads to the parastrophe orientation of the plastids. The movement of the plastids is controlled and mediated by the cytoplasm. The plastids are passively moved by the cytoplasm; therefore the statement "plastids move in the cell" is not correct. The filamentous green alga *Mougeotia* (Fig. 38), which contains only one plate-like chloroplast per cell, is unique in that quite separate photochemical processes seem to control the low intensity and the high intensity

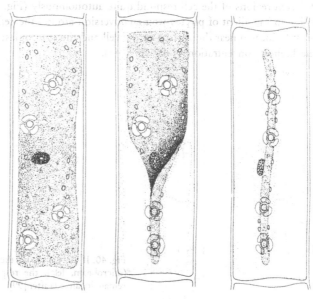

Fig. 38. Individual cells of the filamentous green alga *Mougeotia*. The cylindrical cells contain a single flat plate-like chloroplast. Left, diastrophe ("face position"); right, parastrophe ("profile position"); middle, a transient situation. Note that the light is given from above (after [194])

movement. The low intensity movement in any case is mediated through the phytochrome system.

Filaments of *Mougeotia* are made up of long cylindrical cells each of which contains a single large flat plate-like chloroplast. The chloroplast is able to turn inside the cell. The light-induced movement of this chloroplast is a response to light absorbed by the cytoplasm and not by the chloroplast itself. The low intensity movement of the chloroplast is a typical phytochrome response, as can be shown by the usual simple experiments with consecutive irradiations with small doses of red and far-red light (Fig. 39). Movement induced by a minute of red light can be cancelled by a subsequent treatment with 1 min of far-red light, etc. The movement itself takes about 30 min for completion with a lag-phase of a few minutes.

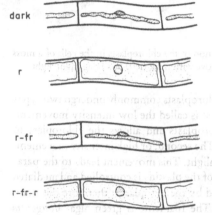

Fig. 39. An experiment which demonstrates the involvement of phytochrome in the low intensity movement (profile position → face position) of the *Mougeotia* chloroplast. Dark, starting position; r, r – fr, r – fr – r, orientation of the chloroplast about 30 min after brief irradiations (1 min each) with red and far-red light (after [99])

Using a microbeam for partial irradiation of the cell it was possible to show that the different regions of the cell respond quite autonomously (Fig. 40), and that the response is due to a gradient of phytochrome conversion around the cell. The active phytochrome must be located near the surface of the cell and the chloroplast always rotates away from the highest concentration of P_{fr} (Fig. 41).

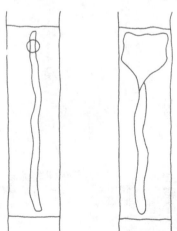

Fig. 40. If only part of the *Mougeotia* cell is irradiated (microbeam, left), the response of the chloroplast will remain localized (after [99])

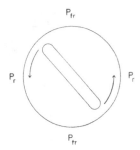

Fig. 41. Model of a cross-section of a *Mougeotia* cell, indicating the direction of chloroplast rotation under conditions of unequal distribution of P_{fr} in the cell. The chloroplast always rotates away from the highest concentration of P_{fr} (after [99])

It turned out that in *Mougeotia* a strong action dichroism can be demonstrated. In linearly polarized red light, given from above, the only cells of the filament to respond with a chloroplast movement are those which are oriented normally to the electrical vector of the linearly polarized light (Fig. 42). This observation points to a strong dichroism of phytochrome, i.e. to a rather precise orientation of dichroic photoreceptor molecules in the neighborhood of the cell surface.

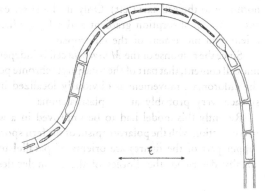

Fig. 42. The behavior of a filament of *Mougeotia* after a brief irradiation with linearly polarized red light from above. The plane of the electrical vector is indicated by the arrow (after [99])

In an attempt to explain the detailed experimental results HAUPT (in 1962) proposed a model (Fig. 43) in which the axis of main absorption (or the dipoles) of the P_r molecules are oriented parallel to the cell surface but obliquely to the cell axis, thus forming a screw around the cell. Of course, each of these dipoles has absorption vectors both parallel and normal to the cell axis, and it can easily be understood (Fig. 44) that under these conditions

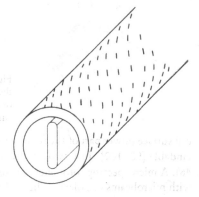

Fig. 43. A model of the *Mougeotia* cell which indicates the dichroic orientation of the phytochrome (P_r) molecules. The axis of main absorption is oriented parallel to the cell surface but obliquely to the cell axis (after [99])

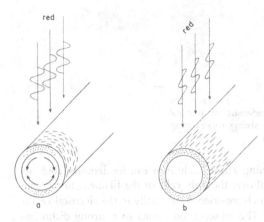

Fig. 44. Models of *Mougeotia* cells (chloroplasts omitted) with the obliquely oriented dichroic phytochrome split into the perpendicular (a) and parallel (b) vectors. Irradiation with linearly polarized light is applied with the electrical vector perpendicular (a) or parallel (b) to the cell axis. The absorption probability is indicated by the dots. The result is an unequal (a) or a homogeneous (b) distribution of P_{fr}. The rotation response of the chloroplast is indicated by the arrows (after [101])

polarized red light is absorbed by the P_r molecules around the whole cell if vibrating parallel to the cell axis (right) but that it is absorbed only at front and rear if vibrating normally to the cell axis (left). Only in the latter case does differential absorption occur, and hence an absorption gradient and a P_{fr}-gradient will arise which will determine the orientation movement of the chloroplast.

Since the response of the *Mougeotia* cell is independent of displacements of the cytoplasmic cell content, that part of the total phytochrome population which is active in determining the chloroplast movement is obviously localized in a rigid dichroic structure at the cell surface, very probably at the plasmalemma.

Recently this model had to be improved in a way which has already been discussed in connection with the polarotropism of the fern sporelings (Fig. 45). While the P_r molecules (upper part of the figure) are orientated parallel to the cell surface in a manner I have already described, the dipoles of the P_{fr} molecules must be orientated normally to the

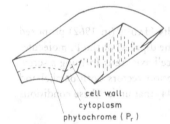

cell wall
cytoplasm
phytochrome (P_r)

phytochrome (P_{fr})

Fig. 45. This model of part of a *Mougeotia* cell represents the orientation of P_r (above) and P_{fr} (below). While P_r is oriented parallel to the surface, P_{fr} is oriented normal to the surface of the cell (after [99])

cell surface (lower part of the figure). Otherwise the new experimental facts are not understandable [98, 100]. Just two of the decisive experiments will be discussed briefly (Fig. 46). A microspectrophotometer was used to irradiate small portions of the *Mougeotia* cell with microbeams of polarized light. The behavior of the chloroplast was used as an indication of the form of the phytochrome. This is possible because the edge of the chloroplast moves so as to avoid that part of the cell where the concentration of P_{fr} molecules is greatest, and only a small part of the chloroplast moves in response to irradiation of a small area of the cell.

The first experiment (upper two lines of Fig. 46): After virtually all the phytochrome in the cell had been established as P_r by an initial saturating dose of far-red light, a part of the edge of the cell was irradiated with a microbeam of polarized red light. Movement of the chloroplast occurred only when the plane of vibration of the light was parallel to the long axis of the cell indicating that the P_r form of the molecule is oriented parallel to the surface of the cell.

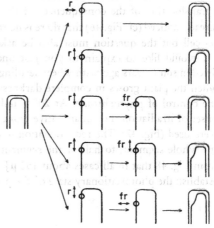

Fig. 46. Microbeam experiments with polarized red (r), followed by far-red (fr) light. Left and middle, situation during irradiation; right, response (after [102])

The second experiment (lower three lines of Fig. 46): The orientation of the P_{fr} form was revealed by following an inductive red irradiation with polarized far-red light. Chloroplast movement was cancelled only when the plane of vibration of the linearly polarized far-red light was at right angles to the long axis of the cell. This indicates that the dipole axis of the P_{fr} form of phytochrome is perpendicular rather than parallel to the surface of the cell.

To summarize: a random orientation of photoreceptors *in vivo* is improbable. In order to understand the excitation and action of biologically significant photoreceptors a knowledge of the state of orientation *in situ* is essential. The best method which has been developed so far is the use of linearly polarized light within the theoretical concept of dichroism.

4. *The situation in higher plants.* It is obvious that most cells do not show physiological responses which could be used to detect dichroism of phytochrome. In the tissue of a leaf, for instance, it is nearly impossible to apply these methods.

Therefore in higher plants the problem of intracellular localization of phytochrome is far from being solved. However, some recent findings indicate that the problem can be approached experimentally. MARMÉ and SCHÄFER [152] reported evidence for membrane-bound phytochrome in maize coleoptiles. Segments of dark-grown maize coleoptiles were irradiated with linearly polarized red or far-red light parallel or normal to the longitudinal axis of the coleoptile. It was found that a non-saturating dose of normally vibrating polarized light converts 20 per cent more P_r to P_{fr} as compared to the same dose of parallel polarized light. This was taken as evidence of phytochrome dichroism along the plasmalemma. In addition, it was found in homogenates of dark-grown maize coleoptiles [151] that a large percentage of phytochrome is "bound" to particles other than nuclei, mitochondria or plastids.

On the other hand, PRATT and COLEMAN [209] who used an immunocytochemical assay which localizes phytochrome *in situ* found indications that within an etiolated oat cell containing phytochrome, the chromoprotein is always associated with both nuclei and plastids, in addition to the cytoplasm. This general intracellular pattern of phytochrome distribution did not change after exposure to light, even though some 90 per cent of the spectrophotometrically detectable phytochrome was lost.

The next question: can phytochrome as such move in the plant tissue, or is there only a translocation of the consequences of the primary reaction of phytochrome? We have already noticed (cf. Fig. 46) that there is no translocation of phytochrome within a *Mougeotia* cell but the question must also be asked about the tissue of a higher plant.

I would like to explain to you just one experiment which was done by WAGNÉ in Sweden some years ago with wheat seedlings [282]. Leaves of a wheat seedling are folded when the plant grows in complete darkness (Fig. 47). The process of unfolding is under the control of phytochrome. At 20° C unfolding is completed about 24 hours after the onset of irradiation. For quantitative work segments 2 cm long of etiolated wheat leaves were used (Fig. 47). The red irradiation – a brief irradiation only – was administered to the whole segment, to half of the segment or to a quarter of the segment. It is evident from Fig. 48 that in all cases about 150 μJ · cm^{-2} are sufficient for saturation, that is, to establish the photostationary state of the phytochrome system characteristic of the wave-

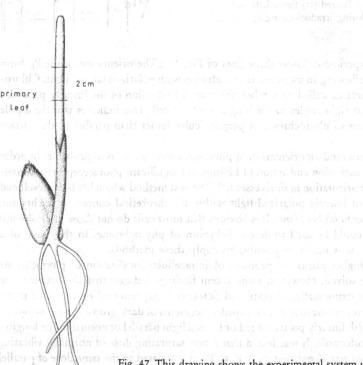

Fig. 47. This drawing shows the experimental system used by WAGNÉ in his experiments: segments, 2 cm long, of etiolated, folded wheat leaves.

Fig. 48. Dose-response curves for induction of wheat leaf unfolding by red light (660 nm). Measurements were taken 24 h after the brief irradiation. Irradiation was administered to the whole leaf segment (upper curve), to half of the segment (middle curve) or to a quarter of the segment (lower curve) (after [282])

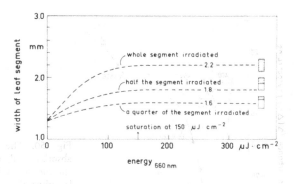

Fig. 49. An illustration of the conclusions drawn from the previous figure (for a detailed description see text) (modified after [282])

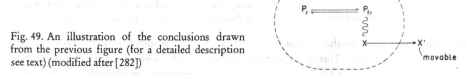

length used, 660 nm ($P_{fr}/P_{tot} \approx 0.8$). In all cases the segment responds as a unit, i.e. no local effect of the red irradiation can be detected. The rate and degree of unfolding is the same throughout the segment. The maximum degree of unfolding, however, depends on how much of the segment has been irradiated. From this type of experiment one has to draw the conclusion (Fig. 49) that both forms of phytochrome and the hypothetical factor X on which P_{fr} primarily acts in the cell are not movable in the wheat leaf tissue. If P_r were freely movable in the tissue most of it would have been converted after a while even when only part of the segment was irradiated. Maximum unfolding should occur when the time of irradiation is prolonged. This requirement is clearly not fulfilled. If factor X were freely movable one could also expect maximum unfolding under all three conditions of irradiation. On the other hand, X', a primary product of the action of P_{fr}, must be translocated rapidly and homogeneously in the wheat leaf tissue since in all cases the segment responds as a physiological unit.

In conclusion, it is probable that phytochrome molecules are always "bound" to certain (possibly several and different) constituents of the cell. So far the interaction cannot be described in molecular terms, but some progress has been made recently in BRIGGS' laboratory at Harvard [64]. The transformation difference spectrum for phytochrome (that is, P_r spectrum minus P_{fr} spectrum) was determined in pea tissue below 560 nm and compared with corresponding difference spectra for extracted pea phytochrome in solution (Fig. 50). The comparison shows that the peak in the difference spectrum occurring in the blue range of the spectrum is shifted to shorter wavelengths and is much enhanced when phytochrome is extracted from the cell and placed in solution. This result clearly indicates that the physico-chemical state of phytochrome in the cell may be different from that of the extracted pigment. This is supported by the observation that there is a shift of the absorption peak of P_{fr}, on extraction, from 735 to 725 nm [64].

These data bear on the so-called "blue problem" which has arisen over the years even in connection with short-term irradiations. In many physiological responses mediated by

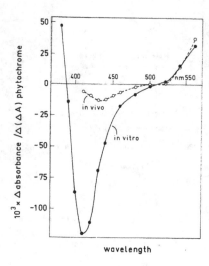

Fig. 50. The transformation difference spectra, absorbance P_r minus P_{fr}, for extracted pea phytochrome in pH 7.8 phosphate buffer (solid line) and for phytochrome in pea hook tissue (broken line). The change *in vivo* is about 10 times smaller and the wavelength of the minimum in the spectrum is shifted from about 430 nm *in vivo* to 413 nm *in vitro*. Despite these changes the signal size at 560 nm is about the same in both spectra (after [64])

phytochrome it has been shown that blue light can cause reactions similar to those caused by red and far-red light. This is understandable since both pigment forms absorb in the blue spectral range (cf. Fig. 9). The action spectrum for phytochrome phototransformation *in vivo* (in corn) has peaks for conversion of P_r to P_{fr} and of P_{fr} to P_r at about 400 nm [206]. Further it shows that red light is 100 times more efficient in converting P_r and far-red light is 25 times more efficient in converting P_{fr} than is blue light. The action spectrum for phytochrome phototransformation *in vitro* (cf. Fig. 8) has peaks in a similar position but red light is only about 6 times more effective in converting P_r and far-red light is only about 4 times more effective in converting P_{fr} than is blue light. Data like the ones shown in Fig. 50 explain the greater effectiveness of blue light *in vitro*. Furthermore, this information is of considerable importance for any quantitative consideration of the phytochrome system which is based on *in vitro* data [94].

It must be emphasized [64] that the *in vitro* spectra in initial crude extracts and increasingly purer preparations of phytochrome are all the same and never have the *in vivo* characteristics; further, altering the environment of extracted phytochrome by changing viscosity, pH, ionic strength, or ionic compositon does not produce a form of phytochrome with the *in vivo* spectral characteristics. Therefore it must be concluded that the *in vivo* spectral properties of phytochrome are the result of a very specific interaction of phytochrome with certain sites in the cell.

Selected Further Reading

OSTER, G.E.: Birefringence and dichroism. In: Physical Techniques in Biological Research, Vol. 1. New York: Academic Press 1955.

ETZOLD, H.: Der Polarotropismus und Phototropismus der Chloronemen von *Dryopteris filix-mas* (L.) Schott. Planta 64, 254 (1965).

JAFFE, L.: Tropistic responses of zygotes of the *Fucaceae* to polarized light. Exp. Cell Res. *15*, 282 (1958).

HAUPT, W.: Localization of phytochrome in the cell. Physiol. Vég. *8*, 551 (1970).

Induction Experiments *Versus* Steady State Experiments; the Problem of the "High Irradiance Response" (HIR)

This is a basic problem from a theoretical as well as from a practical point of view. It implies the question of how phytochrome may function under natural conditions of illumination.

1. Some Phenomena

In the 1ˢᵗ lecture we used light-mediated anthocyanin synthesis in the mustard seedling as an example of a phytochrome-mediated photoresponse. We remember (cf. Fig. 11) that the operational criteria for the involvement of phytochrome in this response are perfectly fulfilled if we use "short-term irradiations" to establish photostationary states ("induction experiments"). Fig. 11 shows several conspicuous features: the kinetics of anthocyanin accumulation are sigmoid after an initial lag-phase of 3 h; the rate of anthocyanin accumulation begins to decrease after about 12 h after irradiation. This latter fact is to be expected since the P_{fr}, originally created by the brief light treatment at time zero, will disappear from the system by irreversible destruction.

What are the kinetics of anthocyanin accumulation if we irradiate *continuously* with standard far-red light, i. e. with a light source which will maintain a nearly stationary concentration of P_{fr} in the mustard seedling over a considerable period of time (cf. Fig. 28)? The answer is given by Fig. 51. We see that even under continuous far-red light the initial

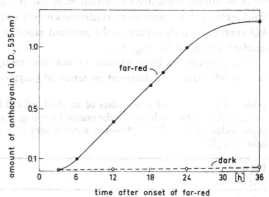

Fig. 51. Time-course of anthocyanin accumulation in mustard seedlings under continuous standard far-red light. The initial lag-phase of the response is about 3 h. Onset of far-red light: 36 h after sowing (after [162])

lag-phase is 3 h, and that a constant rate of synthesis is maintained over at least 24 h. So far the data of Fig. 51 are consistent with our expectations. However, if we vary the irradiance of our standard far-red source we find that the rate of anthocyanin synthesis is a nearly logarithmic function of the quantum flux density over a wide range (Fig. 52). This is unexpected because the photostationary state of the phytochrome system (P_{fr}/P_{tot})

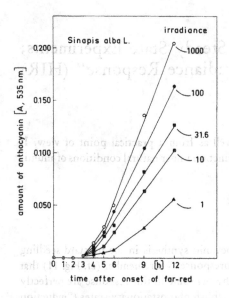

Fig. 52. Initial lag-phase and time-courses of anthocyanin accumulation in mustard seedlings under continuous far-red light as a function of irradiance. Irradiance 1000 denotes the standard far-red irradiance (350 μW $\cdot$ cm^{-2}). Onset of far-red light: 36 h after sowing (after [145])

is a function of wavelength but not a function of irradiance once the photostationary state is established. The serious problem created by these findings has been named the problem of the HIR ("high irradiance response"). The question has been whether or not the HIR can be fully explained on the basis of phytochrome.

The HIR is by no means restricted to anthocyanin synthesis. It has been detected in a great many other responses, two of which will be described briefly.

a) Ascorbate oxidase (AO) is an interesting terminal oxidase of widespread distribution in the plant kingdom, including the mustard seedling. In the dark, enzyme formation is only modest; however, the enzyme can be induced to some extent by phytochrome using the conventional inductive irradiation with red and far-red light (Table 1). It is obvious that the effect of the short-term irradiations is very small as compared with the tremendous AO synthesis which occurs in the mustard seedling under the influence of continuous standard far-red light (Fig. 53).

The problem has been whether or not this strong effect of the continuous standard far-red light can be understood in terms of phytochrome.

Table 1. The influence of short pulses of standard red and far-red light on the increase of ascorbate oxidase (AO) in the cotyledons of the mustard seedling. The operational criteria for the involvement of phytochrome are fulfilled. However, note the very strong "inductive" effect of continuous far-red light (after [52])

Treatment after Sowing	Enzyme Activity [AO/pair of cotyledons]
48 h d[a]	100%
48 h d + 24 h d	149 ± 2,8%
48 h d + 5 min red + 24 h d	200 ± 5.6%
48 h d + 5 min far-red + 24 h d	179 ± 4.2%
48 h d + 5 min red + 5 min far-red + 24 h d	183 ± 5.6%
48 h d + 24 h far-red	380 ± 5.6%

[a] d = dark

b) Phenylalanine ammonia-lyase (PAL) is another enzyme which has been thoroughly investigated in connection with phytochrome-mediated photomorphogenesis of the mustard seedling during the last 6 years [49]. This enzyme catalyzes the formation of trans-cinnamic acid from phenylalanine, one of the most important reactions in the secondary metabolism of higher plants. In the dark-grown seedling only a very small amount of PAL can be isolated from the cotyledons. By a single brief red or far-red irradiation some enzyme synthesis can be induced. This effect may be fully attributed to the traditional phytochrome system, since the operational criteria are fulfilled (Table 2). However, under continuous far-red light the induction effect is very much stronger and the usual irradiance dependency of the HIR is observed (Fig. 54).

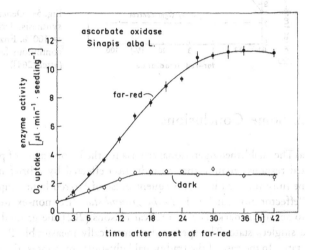

Fig. 53. Time-course of "induction" of ascorbate oxidase in the mustard seedling under continuous standard far-red light. Onset of far-red light: 36 h after sowing (after [203])

Table 2. Reversion experiments which demonstrate that the operational criteria for the involvement of phytochrome in light-mediated "induction" of phenylalanine ammonia-lyase (PAL) are fulfilled (after [264])

Light Treatment	PAL Activity pmoles trans-cinnamic acid / min · pair of cotyledons
t = 0 = time of sowing	
48 h dark	4.1 ± 0.3
52 h dark	5.2 ± 0.5
48 h dark + 5 min red + 4 h dark	13.6 ± 0.6
48 h dark + 5 min far-red + 4 h dark	10.8 ± 0.6
48 h dark + 5 min red + 5 min far-red + 4 h dark	10.4 ± 0.6
48 h dark + 5 min far-red + 5 min red + 4 h dark	14.3 ± 0.8

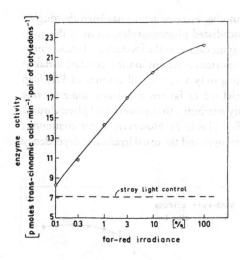

Fig. 54. Dependence on irradiance of PAL synthesis. Irradiance of standard far-red light = 100%. Enzyme extraction: 3 h after onset of continuous far-red light (= 51 h after sowing) (after [264])

2. Some Conclusions

a) The established operational criteria for the involvement of phytochrome in a light-mediated response require that an induction effected by a brief irradiation with red light can be fully reversed by a subsequent pulse of far-red light. Implicitly it is assumed that the "effector molecule" is P_{fr} *in the ground state* (i.e. non-excited state) since P_{fr} is supposed to act in the dark. For theoretical reasons [202] the ground state of P_{fr} is very probably a singlet$_0$ state. The spectrophotometrically measurable P_{fr} represents P_{fr} in the ground state. In the case of the traditional light-pulse responses, the general function which relates the extent of a response, Δm, to the amount of spectrophotometrically detectable P_{fr} must be written as

$$\Delta m = f \; ([P_{fr}] \; \text{ground state})$$

where $[P_{fr}]$ ground state is the concentration of P_{fr} present immediately after termination of a brief irradiation. In principle this function can be elaborated empirically for every individual photoresponse.

In this connection it is necessary to touch briefly on a problem which has been discussed in the previous literature (e.g. [109]) under the term "phytochrome paradoxes". This term designates "non-rational" relationships between the spectrophotometrically detectable phytochrome content and the rate or degree of the particular photoresponse. The so-called "maize paradox" (a representative example) was described in 1966 by BRIGGS and CHON [25]. The phototropic curvature of corn coleoptiles (which itself can only be elicited by blue light) is sensitive to a brief red light treatment, and the sensitizing red light effect is fully reversible with far-red light. However, the red light effect is saturated before any phytochrome transformation can be detected with the standard "Ratiospect". An appropriate explanation requires the assumption that a small phytochrome population exists which mediates this sensitizing effect and which is characterized by very high photochemical rate constants (i.e. very short half-lives) for both photochemical reactions, $P_r \rightarrow P_{fr}$ and $P_{fr} \rightarrow P_r$. We remember from the 2nd lecture that Schäfer and associates [150, 249] have indeed observed signals which possibly indicate that several phytochrome populations exist *in vivo* which differ greatly with respect to their photochemical rate constants.

b) For good reasons the HIR can also be attributed to the phytochrome system [94, 264]. The irradiance dependency of the HIR has been explained by the hypothesis that P_{fr} can act from some excited state, P_{fr}^*, which is much more physiologically active than P_{fr} in the ground state. Thus, we explicitly introduce a second species of P_{fr} (besides P_{fr} in the ground state) which is physiologically effective, P_{fr}^*. While speculations about the nature of the excited state have been advanced [94], at present this state cannot be measured directly, i.e. by physical means. Therefore it is not possible at present to elaborate the function

$$\Delta m = f ([P_{fr}^*])$$

wherein Δm is the extent of the response (in general a rate) and $[P_{fr}^*]$ is a steady-state concentration of the excited species. In any case $[P_{fr}^*]$ is a function of irradiance irrespective of the nature of the excited state. When the light is turned on P_{fr}^* appears rapidly; at the moment when the light is turned off the excited species disappears rapidly, on the order of 10^{-10} to 10^{-4} s, depending on the physical nature of the excited species.

3. Control of Hypocotyl Lengthening in Lettuce (*Lactuca sativa* L., cv. Grand Rapids) as a Prototype of a "High Irradiance Response" [94, 97]

Hypocotyl lengthening in lettuce is a very convenient response to study the HIR: it is a "negative" response (inhibition); it is not significantly controlled by P_{fr} in the ground state; hypocotyl lengthening is exclusively due to cellular lengthening during the experimental period [88].

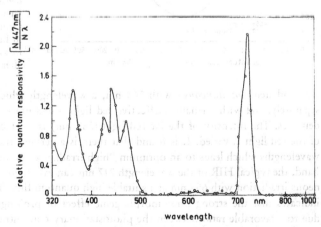

Fig. 55. An action spectrum for control (inhibition) of hypocotyl lengthening in lettuce seedlings (*Lactuca sativa*, cv. Grand Rapids) by continuous light. The action spectrum was elaborated between 54 and 72 h after sowing. During this period lengthening of the hypocotyl is almost exclusively due to cell lengthening. Note that this action spectrum is *not* for the "induction" of a response where the reciprocity law is valid (cf. Fig. 5) but for a "long-term" irradiation where reciprocity does not hold. However, a steady state situation (i. e. a constant rate of the response) can probably be assumed for all wavelengths during the period of light treatment (supplemented after [97])

I shall describe briefly those experimental results (obtained by K.M. HARTMANN) which have proven *experimentally* (or operationally) that the HIR is due to phytochrome in some way or other. Even if some of HARTMANN's *theoretical* concepts may not survive long, the experimental evidence remains.

Let us recall the action spectrum for a phytochrome-dependent response in the case of "short-term" irradiations, i.e. at a maximum 5 min or thereabouts of irradiation (cf. Fig. 5). This action spectrum reflects the effectiveness of the different wavelengths in transforming P_r into $P_{fr \text{ (ground state)}}$. Fig. 55 represents the action spectrum for the control of hypocotyl lengthening in lettuce seedlings by continuous ("long-term") light over an 18 h light period, 54 to 72 h after sowing at 25 °C. It is obvious that the "long-term" action spectrum is totally different from the "short-term" action spectrum. If we exclude the blue part of the spectrum (< 520 nm) from our present considerations, we realize that quanta between 500 and 700 nm are barely effective and that far-red light above 750 nm is totally ineffective. By contrast, we observe a very strong and nearly symmetric band of action close to 720 nm where the photoequilibrium of the phytochrome system contains only about 3 per cent P_{fr} (cf. Fig. 10). The problem is whether or not the far-red peak of action can be understood on the basis of the phytochrome system.

A successful approach to solving this problem is illustrated in Fig. 56. Lettuce seedlings

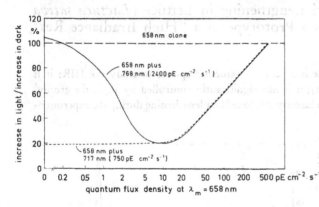

Fig. 56. Hypocotyl lengthening in lettuce seedlings (*Lactuca sativa*, cv. Grand Rapids) under continuous simultaneous irradiation with different quantum flux densities in the red (658 nm) at a constant high quantum flux density background of different far-red bands (717 nm, 768 nm). Ordinate: ratio between "increase in light" and "increase in dark", measured between 54 and 72 h after sowing (after [96])

are irradiated *simultaneously* with 768 nm, a wavelength which is ineffective if applied separately, and with virtually ineffective red light – 658 nm – at different quantum flux densities. The intensity of the far-red light (768 nm) is kept constant and the intensity of the red light is varied. It is found that there is a certain irradiance ratio of these two wavelengths which leads to an optimum "high irradiance response" (HIR). On the other hand, the typical HIR of the wavelength 717 nm can eventually be nullified by a simultaneous irradiation with 658 nm at a suitable high quantum flux density. This type of result indicates that the strong photomorphogenic effect of prolonged far-red light might be due to a favorable ratio between the photostationary concentrations of P_r and P_{fr}. Using the same experimental approach with other pairs of wavelengths, it was found that a maximum "high irradiance response" with respect to hypocotyl lengthening will always occur when the ratio $[P_{fr}] / [P_{tot}]$ is about 0.03. If the ratio $[P_{fr}] / [P_{tot}]$ is above 0.3 or below 0.002 the action of light on hypocotyl lengthening shows less than 10 per cent deviation from the dark control.

Two critical questions arise immediately:

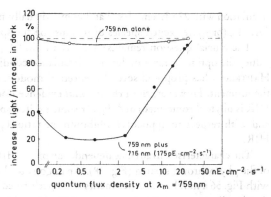

Fig. 57. Hypocotyl lengthening in lettuce seedlings (*Lactuca sativa*, cv. Grand Rapids) under continuous simultaneous irradiation with different quantum flux densities at 759 nm at a constant quantum flux density background of the wavelength 716 nm. Ordinate: ratio between "increase in light" and "increase in dark", measured between 54 and 72 h after sowing (after [96])

a) Is P_{fr} (or some excited species of P_{fr}) the effector molecule of the "high irradiance response"? If P_{fr} (or some excited species of P_{fr}) is the effector proper one can predict that the far-red band of action (around 720 nm) can be nullified through a simultaneous irradiation with a high irradiance black-red band (759 nm) which is almost exclusively absorbed by P_{fr} and which therefore depresses the steady state concentration of P_{fr} to close to zero.

The experimental result is shown in Fig. 57. It is found that a quantum flux density of 175 pE $\cdot$ cm^{-2} $\cdot$ s^{-1} at 176 nm depresses hypocotyl lengthening by about 60 per cent. On the other hand the wavelength 759 nm is virtually ineffective up to 28 nE$\cdot$cm^{-2}$\cdot$s^{-1}. It is also found that simultaneous irradiation with 175 pE$\cdot$cm^{-2}$\cdot$s^{-1} at 716 nm and, e.g., 300 pE$\cdot$cm^{-2}$\cdot$s^{-1} at 759 nm depresses hypocotyl lengthening by 80%. This fact indicates that the photoequilibrium at 716 nm is somewhat higher than the optimum value. Maximum action is expected to be close to 720 nm. Indeed, the peak of the action spectrum is located at 720 nm (cf. Fig. 55). The other important feature of Fig. 57 is that the simultaneous

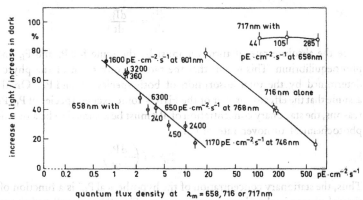

Fig. 58. Hypocotyl lengthening in lettuce seedlings (*Lactuca sativa*, cv. Grand Rapids). Ordinate: ratio between "increase in light" and "increase in dark", measured between 54 and 72 h after sowing. Left curve, the "ratios" for response maxima taken from curves like the one in Fig. 56 are plotted as a function of the quantum flux density of the red light (658 nm). Remember from Fig. 56 that the red light is applied simultaneously with a constant high quantum flux density background of different far-red bands (801 nm, 768 nm, 746 nm). The parallel curve on the right is a dose-response curve for 716 nm only. The upper curve (at the 90 per cent level) is based on data obtained with simultaneous irradiation with 717 nm and 658 nm. This curve shows that the 90 per cent "ratio" is established when the ration P_{fr}/P_{tot} is around 0.3 (after [96])

irradiation with 27 nE·cm⁻²·s⁻¹ at 759 nm virtually nullifies the action of 716 nm. Therefore, P_{fr} (or some excited species of it) must be the specific effector molecule of the HIR.

The crucial question remains: Why is a ratio $[P_{fr}]$ / $[P_{tot}]$ = 0.03 (or a slightly lower
value) the optimum ratio under high irradiance conditions? To account for this problem
HARTMANN has proposed several theoretical models, none of which is fully conclusive at
the moment. However, the *experimental* evidence does seem to be fully convincing: the
HIR is phytochrome-mediated, P_{fr} (or some excited species of P_{fr}) is the effector molecule,
and with respect to hypocotyl inhibition the ratio $[P_{fr}]$ / $[P_{tot}]$ ≈ 0.03 is optimum for the
HIR.

The characteristic irradiance dependency of the HIR can also be attributed to properties
of the phytochrome system (Fig. 58). Analogous to the procedure described in connection
with Fig. 56 a number of experiments were performed in which red light (658 nm) of increasing irradiance was applied together with a far-red wavelength at constant irradiance. In
this way response maxima were determined with a number of different wavelengths as
the second partner of 658 nm (e.g. 658/801, 658/768, 658/746). These far-red wavelengths
are almost exclusively absorbed by P_{fr}. While the photoequilibrium for the response maxima
is always $[P_{fr}]$ / $[P_{tot}]$ ≈ 0.03, the degree of the effect obtained at the maximum is very different, ranging from about 20 up to about 80 per cent inhibition. This is a consequence
of the irradiance dependency. Namely, if we plot the "ratio" [increase in light/increase
in dark] as a function of the quantum flux density which is required at 658 nm for an
optimum response with the different partner wavelengths, a dose-response curve is found
which parallels the characteristic "high irradiance" dose-response curve for the wavelength
716 nm. This fact clearly shows that the *total absorption of quanta* in the phytochrome
system will determine the extent of the HIR once a certain photoequilibrium is established.

Theoretically, this experimental result can be expected. The photoequilibrium of the
phytochrome system can be defined as

$$-\frac{dP_r}{dt} = -\frac{dP_{fr}}{dt}$$

since the absolute rate of turnover must be the same for P_r and P_{fr} under conditions of
photoequilibrium. This means that the rate of conversion of the phytochrome system is
determined by the total absorption of both species, P_r and P_{fr}. On the other hand we
assume that the effector proper of the HIR is some excited species of P_{fr}, P_{fr}^*. For theoretical
reasons, the stationary concentration of P_{fr}^* must be assumed to be a function of the absolute
photochemical turnover rate of P_{fr}.

$$P_{fr}^* = f\left(-\frac{dP_{fr}}{dt}\right)$$

Thus, the stationary concentration of the hypothetical P_{fr}^* is a function of the total absorption of quanta in the phytochrome system. The "long-term" action spectrum of Fig. 55
does not reflect the absorption spectrum of a single photoreceptor but results from the
simultaneous excitation of two different photoreceptors (P_r and P_{fr}).

In conclusion, we must discriminate between P_{fr} (ground state) and $\bar{P}_{fr}^*$ and ascribe
at least the major part of the effect of continuous standard far-red light to the action of
P_{fr}^*. Unfortunately, the interpretation of P_{fr}^* in physical terms is still an open question;
however, the *experimental* evidence that the action of continuous far-red light is due in
some way to P_{fr} has been convincingly presented by K.M. HARTMANN [94, 95, 96, 97].

4. Further Applications of HARTMANN's Technique

HARTMANN's technique (simultaneous irradiation with two wavelengths) has been successfully used by other investigators. A first example: BLONDON and JACQUES [16] have studied the action spectrum of flower initiation in the long-day plant *Lolium temulentum*. A sharp peak of action was found between 700 and 725 nm if only monochromatic light was used. However, if the monochromatic light was applied in addition to a background irradiation with red light, the peak of action was shifted towards longer wavelengths ($\lambda_{max} > 725$ nm). In both cases the maximum of action corresponded to the same photostationary state of the phytochrome system.

A second example: HACHTEL [86] has used the technique of simultaneous irradiation with two wavelengths to demonstrate that anthocyanin formation in *Oenothera* seedlings under continuous far-red light is mediated by phytochrome (Fig. 59).

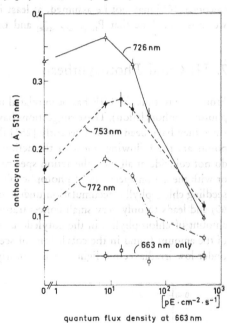

Fig. 59. Formation of anthocyanin in *Oenothera* seedlings under continuous simultaneous irradiation with different quantum flux densities at 663 nm at a constant quantum flux density background of the wavelengths 726 nm (528 pE · cm^{-2} · s^{-1}), 753 nm (548 pE · cm^{-2} · s^{-1}), and 772 nm (562 pE · cm^{-2} · s^{-1}). The amount of anthocyanin was determined 60 h after the onset of irradiation (after [86])[1]

5. The Action of Blue Light in "Long-term" Experiments

HARTMANN's interpretation of the far-red band of action of the HIR can scarcely be applied to cases where a strong HIR is observed in the blue part of the spectrum without any action in the far-red range as, for example, for anthocyanin synthesis in Wheatland milo [51]. Furthermore, it seems necessary to assume even in the case of a blue-far-red action spectrum that in the blue spectral range ($\lambda < 520$ nm) a "high irradiance response" comes

---

1 HACHTEL's data are obviously not in agreement with the conclusion (cf. page 42) that the photoequilibrium [P$_{fr}$]/[P$_{tot}$] for the response maxima is always the same. This type of discrepancy indicates that the explanation of the HIR is not yet fully satisfactory.

into play which cannot fully be interpreted on the basis of phytochrome. The detailed action spectrum obtained by HARTMANN for hypocotyl lengthening in lettuce indicates (cf. Fig. 55) that the effective light below 520 nm is predominantly absorbed by a flavoprotein.

6. Sequential Action of "High Irradiance Reaction" and P_{fr} (in the Ground State)?

The initial lag-phase of anthocyanin is exactly the same whether the biosynthetic process is induced by a brief irradiation with red or far-red light whereby the law of reciprocity is valid (cf. Fig. 11), or by continuous far-red light (HIR) (cf. Fig. 52). This fact shows that the sequential action of a "high irradiance reaction" (first) and P_{fr} (in the ground state) (second) [105] may not be assumed, at least in the case of the mustard seedling. Rather, we must conclude that $P_{fr\ (ground\ state)}$ and the P_{fr}^* act on the same system.

7. HIR and Photosynthesis

From time to time the HIR has been related to photosynthesis (or to parts of it like cyclic photophosphorylation). These suggestions have been refuted previously [10, 87]; however, since they have been revived recently [253] they must be explicitly disproved. The main points are the following: most of the action spectra of the HIR (as shown by Fig. 55) do not coincide at all with the action spectra of the two photosystems of photosynthesis or with the action spectrum of photophosphorylation [299, 301, 123, 79]; in the mustard seedling chlorophyll a accumulation under continuous standard far-red light is slow (Fig. 60) and leads to only very small concentrations (36 h after the onset of far-red light the amount of chlorophyll a in the cotyledons of the mustard seedling is only 2.2 per cent of the amount found in the cotyledons of seedlings which were grown for the same time under low irradiance red light which is nearly ineffective as far as the HIR is concerned,

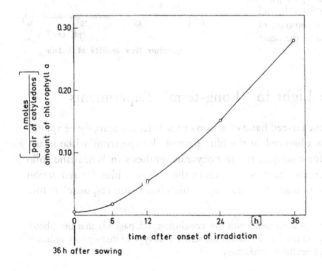

36 h after sowing

Fig. 60. The time-course of chlorophyll a accumulation in the cotyledons of the mustard seedling under continuous standard far-red light. The value at 36 h after onset of far-red light is only 2.2 per cent of the chlorophyll found in cotyledons of seedlings grown under weak red light (after [153])

cf. Fig. 55); when we compare Fig. 52 and Fig. 60 we see that the extent of the HIR is not related to the amount of chlorophyll a. The steady state rates of the "high irradiance responses" of the mustard seedling are established in a matter of 3 h or less (e.g. Fig. 85), and the initial lag-phases of the "high irradiance responses" do not depend on irradiance while the rate of chlorophyll accumulation does, etc.

Unfortunately, there are no data available on cyclic photophosphorylation in the mustard cotyledons. However, data obtained by OELZE and BUTLER [191] for primary leaves of beans indicate that the capacity for photophosphorylation appears only after 12 h in far-red light.

In conclusion, there are no data which indicate a relationship between the HIR and the function of the photosynthetic apparatus.

8. Operational Definitions (Criteria) for the Involvement of Phytochrome in a Response

Involvement of $P_{fr\ (ground\ state)}$: an "induction" performed by a brief pulse with red light is fully reversed by immediately following it with a corresponding pulse of far-red light.

Involvement of P_{fr}*: a "high irradiance response" mediated by continuous near far-red light (e.g. 720 nm) can be obtained by simultaneously irradiating with red light (e.g. 660 nm) and long-wavelength far-red light (e.g. 760 nm) following the experimental procedure described in Fig. 56.

Remember that an increase of the signal $\Delta\Delta A$ is in general called "phytochrome (P_r) synthesis" while a decrease of the signal $\Delta\Delta A$ is in general called "phytochrome (P_{fr})decay".

Selected Further Reading

HARTMANN, K.M.: A general hypothesis to interpret high energy phenomena of photomorphogenesis on the basis of phytochrome. Photochem. Photobiol. 5, 349 (1966).

HENDRICKS, S.B., BORTHWICK, H.A.: The physiological functions of phytochrome. In: Chemistry and Biochemistry of Plant Pigments. (T. W. GOODWIN, edit.). London: Academic Press 1965.

MOHR, H.: Photomorphogenesis. In: The Physiology of Plant Growth and Development (M.B. WILKINS, edit.). London: McGraw-Hill 1969.

ROLLIN, P.: Phytochrome, Photomorphogénèse et Photopériodisme. Paris: Masson 1970.

Phytochrome and the Diversity of Photoresponses; "Positive" and "Negative" Photoresponses; a Unifying Hypothesis

The first four lectures of this course have been used mainly to describe some important properties of phytochrome *per se*. Now let us try to bring the complexity of the living system into our considerations. P_{fr} can be regarded as the physiologically effective form of the phytochrome system. The photochemical formation and maintenance of this specific chromoprotein in a plant will lead to a great number of photoresponses which together constitute the phenomenon of photomorphogenesis (cf. Fig. 1, 2). We must understand "photomorphogenesis" in molecular terms. The experimental material in many instances has been the dark-grown mustard seedling (*Sinapis alba*) between 36 and 72 h after sowing at 25° C (cf. Fig. 1). During this period there is no significant increase of cell number or DNA contents in either organ, cotyledons or hypocotyl [288]. For this reason the biological unit (organ) – pair of cotyledons or hypocotyl – can be used as a system of reference for the enzyme data instead of cell or unit DNA.

Let us recall the conclusion drawn previously (cf. Fig. 28) that under continuous far-red light a low but nearly stationary concentration of the active P_{fr} can be maintained over a considerable period of time. This circumstance is a very favorable one for our studies; for if we investigate the process of photomorphogenesis under continuous far-red light we deal with nearly steady state conditions of the phytochrome system. An additional advantage is that this part of the spectrum exerts the strongest effect on morphogenesis (cf. Fig. 55) and that any interaction of photosynthesis remains excluded.

These points have convinced us [162] that it might be best to study the causalities of photomorphogenesis under continuous far-red light. However, we must be aware that under these conditions the "high irradiance responses" (HIR) will dominate. This means (if we make use of the terminology developed in the fourth lecture) that we are mainly (possibly exclusively) studying the effects exerted by P_{fr}^* on morphogenesis. Continuous far-red light of high irradiance may even satisfy the ecologist. We are close to natural conditions insofar as high irradiance far-red light does fully replace high irradiance white light as far as *photomorphogenesis* is concerned (cf. Fig. 1).

If we desire to understand the phenomenon of photomorphogenesis we must ask the following question: what are the causal relationships between P_{fr} ($P_{fr\ (ground\ state)}$ as well as P_{fr}^*) and the final photoresponses – for example, light-induced growth of the cotyledons of a seedling, or light-induced synthesis of a compound like anthocyanin? To put the question more precisely, let us again consider two mustard seedlings of the same chronological age, a dark-grown seedling on the left and a corresponding light-grown seedling on the right (cf. Fig. 169). All photoresponses of the light-grown seedling are consequences of the formation of P_{fr} in the cells of the seedling.

Some of the many photoresponses are enumerated in Table 3. Let us keep a few in mind, for instance inhibition of hypocotyl lengthening, enlargement of cotyledons, synthesis of anthocyanin and hair formation along the hypocotyl. As we remember we have very good reasons for stating that phytochrome will be the same molecule irrespective

Table 3. Some phytochrome-mediated photoresponses of the mustard seedling, *Sinapis alba* L. (investigations carried out in our laboratory since 1957)

Inhibition of hypocotyl lengthening
Inhibition of translocation from the cotyledons
Enlargement of cotyledons
Unfolding of the lamina of the cotyledons
Hair formation along the hypocotyl
Opening of the hypocotylar ("plumular") hook
Formation of leaf primordia
Differentiation of primary leaves
Increase of negative geotropic reactivity of the hypocotyl
Formation of tracheary elements
Differentiation of stomata in the epidermis of the cotyledons
Formation of plastids in the mesophyll of the cotyledons
Changes in the rate of cell respiration
Synthesis of anthocyanin
Increase in the rate of ascorbic acid synthesis
Increase in the rate of carotenoid synthesis
Increase in the rate of protochlorophyll formation
Increase of RNA synthesis in the cotyledons
Decrease of RNA contents in the hypocotyl
Increase of protein synthesis in the cotyledons
Changes in the rate of degradation of storage fat
Changes in the rate of degradation of storage protein

of the organ or tissue of a plant from which it is isolated. Stated more straightforwardly, phytochrome *as a molecule* seems to be the same in all the cells of the seedling in which it occurs. But as we observe, the different organs and tissues of the seedling respond differently to the formation of P_{fr}. We have been calling this phenomenon the "multiple action" of P_{fr}. By this term we mean that P_{fr} releases in a plant, e.g. in a seedling, a great number of photoresponses more or less simultaneously. The *specificity* of these responses does not depend on P_{fr}. Which response can take place rather depends on the "specific state of differentiation" of the particular cells and tissues at the moment when P_{fr} is formed in the seedling. We cannot overestimate the significance of this observation if we try to study the problem of what the causalities are which govern the processes between P_{fr} and the final photoresponses. This point is emphasized in Fig. 61. This shows segments of cross-sections through the hypocotyl of mustard seedlings, on the left from a dark-grown seedling, on the right from a seedling of the same age which has been kept in continuous far-red light for some time. One sees that under the influence of phytochrome (P_{fr}^{*} as well as $P_{fr\ (ground\ state)}$) certain cells of the epidermis have formed long hairs and one further sees that all the cells of the subepidermal layer – but no other cells – have formed anthocyanin. It is evident from this simple drawing that P_{fr} functions only as a "trigger"; the specificity of the response – for example hair formation or anthocyanin synthesis – must depend on the "specific state of differentiation" of the cells and tissues at the moment when P_{fr} is formed in the seedling. We have been calling this specific state of differentiation "primary differentiation (P_{fr})", i.e. primary differentiation with respect to P_{fr}. A synonymous expression would be "competence (P_{fr})", i.e. competence of the cells or tissues with respect to P_{fr}. Note that the term primary differentiation (P_{fr}) or competence (P_{fr}) designates, in connection with photomorphogenesis, the pattern of differentiation of a seedling *before* the first formation of P_{fr}. This primary pattern of differentiation or competence may not be regarded

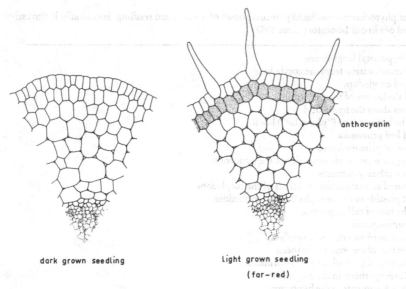

anthocyanin

dark grown seedling   Light grown seedling
                                         (far-red)

Fig. 61. Segments of cross-sections through the hypocotyl of mustard seedlings (*Sinapis alba* L.) grown in the dark or under far-red light (after [285])

as static; rather it is dynamic, that is, it will change with time. One can show, for example, that the ability of the subepidermal cells of the hypocotyl and the epidermal cells of the cotyledons to respond to the formation of P_{fr} with anthocyanin synthesis will first increase, then decrease and finally disappear with increasing age of the seedling. Fig. 62 shows how rapidly the competence of the cells for P_{fr} will change. The seedlings were in all cases irradiated for 12 hours with far-red light. The time of the onset of light was different as is indicated on the abscissa.

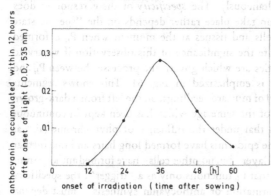

Fig. 62. The amount of anthocyanin accumulated by the mustard seedling within 12 h after the onset of standard far-red light, as a function of seedling's age (after [285])

One can investigate this changing pattern of primary differentiation on a quantitative basis if one divides the seedling, for instance into 4 parts. The type of partition we have been using is sketched in Fig. 63. The radicle can be ignored, as it does not form any anthocyanin. Fig. 64 contains some quantitative data on the amount of anthocyanin which is synthesized in the different segments under the influence of 24 hours of far-red light. The numbers on the bottom indicate the time of onset of the far-red light. I do not want to analyze these data further. They merely illustrate the changing pattern of primary differ-

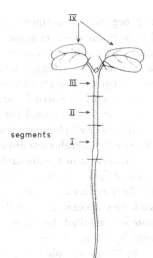

Fig. 63. The mustard seedling is divided into 4 parts, immediately before anthocyanin extraction. The radicle is omitted. The changing potential for the formation of anthocyanin is traced in the different parts (next figure) (after [285])

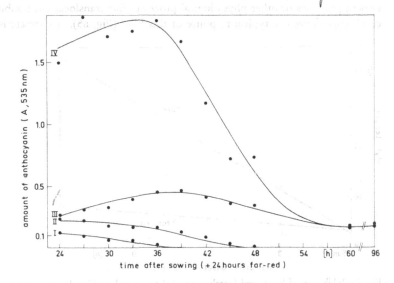

Fig. 64. Ordinate: amount of anthocyanin present in the different segments (cf. Fig. 63) at the end of the irradiation period. Abscissa: time of onset of standard far-red light (hours after sowing). Treatment: the seedlings were irradiated over a period of 24 hours. Immediately afterwards segmentation and extraction were performed (after [285])

entiation. It is evident from these and many other results, that phytochrome research will not immediately offer an opportunity to investigate the causalities of "primary differentiation"; however, we can expect to understand with the help of phytochrome many aspects of "secondary differentiation". Since all hormone effects on development known so far in plants, animals and man are part of secondary differentiation, phytochrome-mediated secondary differentiation may even serve as a logical and molecular model system to investigate hormone effects in general. To approach this problem it turned out to be convenient to divide the many photoresponses of the mustard seedling into 3 distinct categories: positive, negative and complex photoresponses. I shall illustrate each of the three categories by one characteristic example. The pertinent experiments were carried out under carefully controlled conditions. The onset of standard far-red light is always 36 hours after sowing. By this time the seedlings can readily be handled for biochemical and biophysical measurements. Up to 72 hours after sowing development of the mustard seedling is not limited by the supply of internal storage material [112]. There is a period of at least 36 hours which can safely be used for our type of investigation.

a) "Positive" photoresponses are those which are characterized by an initiation or an increase of biosynthetic or growth processes. An example is phytochrome-induced synthesis of anthocyanin in the mustard seedling (cf. Fig. 51). The long initial lag-phase after the onset of light – 3 hours – is characteristic of this type of photoresponse. We remember by the way that continuous far-red light will maintain in the seedling a low but nearly stationary concentration of P_{fr} over a considerable length of time. The photostationary state is established about 1 min after the onset of far-red light. In the dark-grown seedling only P_r will be present.

b) "Negative" photoresponses are those which are characterized by an inhibition of growth processes or other physiological processes like translocation. Inhibition of hypocotyl lengthening is a typical response of this kind (Fig. 65). A characteristic feature of

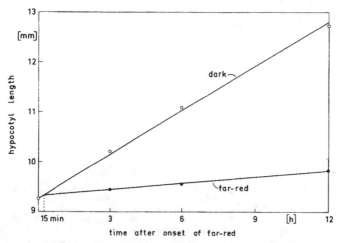

Fig. 65. Inhibition of hypocotyl lengthening of the mustard seedling by continuous standard far-red light. The lag-phase of the response is probably less than 15 min. Onset of light: 36 h after sowing (after [222])

a negative photoresponse is the short lag-phase. In the present case we see that after about 15 min after the onset of far-red light, that is, after the first appearance of P_{fr} in the seedling, a new steady state rate of lengthening is established. The corresponding lag-phase must be very short. Fig. 66 illustrates the principle further. It shows that the rate of translocation of dry matter from the cotyledons into the hypocotyl tissue is reduced with the same short lag-phase as soon as P_{fr} appears in the seedling.

c) Fig. 67 gives an example for the type of response we have been calling "complex" photoresponses. Complex photoresponses are those which are characterized during the first part of the kinetics by an inhibition and later on by a promotion of the response compared with the corresponding dark controls. The regulation by phytochrome of the

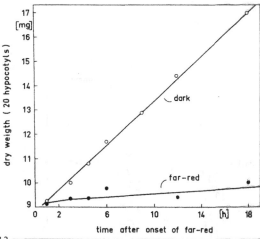

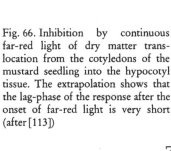

Fig. 66. Inhibition by continuous far-red light of dry matter translocation from the cotyledons of the mustard seedling into the hypocotyl tissue. The extrapolation shows that the lag-phase of the response after the onset of far-red light is very short (after [113])

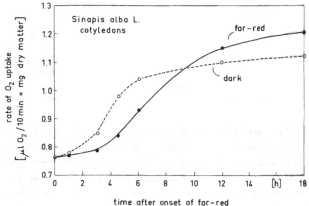

Fig. 67. Control by continuous far-red light of the rate of oxygen uptake by the cotyledons of the mustard seedling. Onset of light: 36 h after sowing (after [112])

rate of O_2 uptake by the cotyledons is an example of this sort of photoresponse. If we follow the kinetics of the rate of O_2 uptake of the cotyledons in darkness and under the influence of continuous far-red light we observe that phytochrome first depresses the rate of respiration; later on, however, far-red light increases the rate of respiration above the corresponding dark control.

It turned out that the complex photoresponses are probably explainable in terms of an interaction of positive and negative photoresponses. As a matter of fact we are left with positive and negative photoresponses which are different in all respects except that both types are mediated by P_{fr}.

The next question has been whether or not P_{fr} can mediate positive as well as negative photoresponses in one and the same cell. To answer this question we look with the microscope at a longitudinal section through the hypocotyl of a mustard seedling (Fig. 68). These drawings represent the three outer cellular layers of the hypocotyl. Left, from a dark-grown seedling and right, from a far-red-grown seedling. We realize that, for example, an epidermal cell can respond to P_{fr} with the formation of a hair, a positive photoresponse, as well as with the reduction of the rate of lengthening, a negative photoresponse. A subepidermal cell can respond with the formation of anthocyanin as well as with the reduction of lengthening.

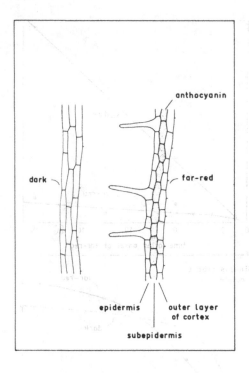

Fig. 68. The drawings represent the three outer cellular layers of the hypocotyl of a mustard seedling. Left, from a dark-grown seedling; right, from a seedling which was kept for some time under continuous far-red light (after drawings by M. Häcker)

The formal interpretation of the situation is as follows: every cell which responds to P_{fr} is competent for P_{fr}. However, the competence is specified: different cells respond differently, and even within one and the same cell different systems respond differently. We must remember one very important detail, i.e. that the initial lag-phase of anthocyanin synthesis is 3 h; however, the corresponding lag-phase for inhibition of lengthening of the *same cell* is less than 15 min. These data are very important if the "mechanism" of the primary reaction of P_{fr} is to be discussed.

Now let us again raise the question of how phytochrome acts. When we ask this question we must be prepared to find that there might be no *general* answer. We recall the "multiple action of P_{fr}" which is plainly illustrated by the magnificent diversity of photoresponses, even within a particular cell. It seems that the question "how does P_{fr} act?" can only be asked in connection with a particular photoresponse (i.e., in connection with a particular system) but not in general.

However, there are good reasons for believing that the hypothesis illustrated in Fig. 69 possibly describes some features of the action of P_{fr} on *morphogenesis* in a developing plant.

It is now generally accepted that all living cells of a particular plant – except perhaps the sieve tube elements – contain the total complement of genetic DNA that is characteristic for the individual; in other words, all the genes are present in all cells but only a fraction of them are active at a given time in a given cell. Genes are thus turned off and on. Such differential gene activity results in differences among cells that have the same set of genes.

In connection with photomorphogenesis we must refine this general scheme in the following way: the total genes of each particular cell of a dark-grown seedling which is able to respond to P_{fr} must be divided into at least four functional types: active, inactive,

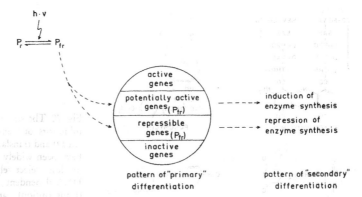

Fig. 69. This scheme illustrates the hypothesis of differential gene activation and differential gene repression through P_{fr}. (The hypothesis is explained in some detail in the text) (modified after [256])

potentially active, and repressible genes. Active genes are those which function the same way in an etiolated plant as they do in a light-grown plant; inactive genes are active neither in the dark-grown seedling nor in the seedling exposed to light (e.g. flowering genes). Potentially active genes with an index P_{fr} are those which are ready to function and whose activity can be started or increased in some way by P_{fr}. The activation of potentially active genes leads to positive photoresponses. Repressible genes with an index P_{fr} are those which can be repressed by P_{fr}. The repression of repressible genes leads to "negative" photoreponses. Which genes are active, inactive, potentially active or repressible in a particular cell at a particular moment is determined by those regulating factors which determine the dynamic pattern of primary differentiation (or competence) in a multicellular seedling. The nature of these regulating factors of primary differentiation is virtually unknown in both plant and animal embryology.

This sort of reasoning is supported by a considerable amount of evidence available at present. If we include the common credo of molecular biology

$$\text{gene (DNA)} \rightarrow \text{RNA} \rightarrow \text{protein}$$

in our reasoning the hypothesis can be written more operationally and can be subjected to experimental verification at the molecular level. Thus, the heuristic value of the hypothesis is obvious even if one is reluctant to accept it as the only explanation of phytochrome action on morphogenesis. There are two principal approaches which can be used to check this hypothesis:

1. Use of specific inhibitors. The results of such experiments are never completely unambiguous and thus must be treated with considerable caution.

2. Direct measurements of P_{fr}-mediated enzyme induction and enzyme repression.

Let us close this lecture with a brief report about the results of inhibitor experiments. In the lectures which follow, we shall deal with enzyme induction and enzyme repression by P_{fr}.

a) Anthocyanin synthesis. You recall that anthocyanin accumulation mediated by far-red light in the mustard seedling is linear with time for at least 20 h after the initial lag-phase (cf. Fig. 51). One may conclude that over this period of time a virtually stationary concentration of P_{fr} (and/or P_{fr}^{*}) will lead to a constant rate of anthocyanin synthesis. If the hypothesis that P_{fr}-mediated anthocyanin synthesis is due to differential gene activation is justified,

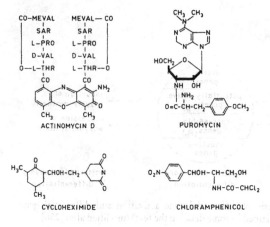

Fig. 70. The structures of some specific inhibitors of transcription (Actinomycin D) and translation. These inhibitors have been widely used to inhibit more or less selectively the processes of DNA-dependent RNA synthesis (transcription) and RNA-dependent protein synthesis (translation)

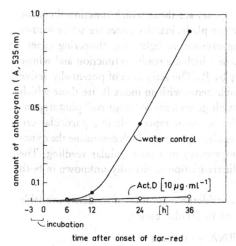

Fig. 71. Time-courses of anthocyanin accumulation in the mustard seedling under the influence of continuous far-red light with and without a pretreatment ("incubation") with Act. D [10 µg · ml⁻¹]. Onset of light: 24 h after sowing (after [170])

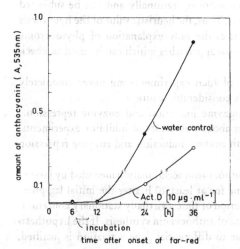

Fig. 72. Time-courses of anthocyanin accumulation in the mustard seedling under the influence of far-red light. Seedlings were irradiated with far-red light for 6 h, transferred to the dark for one hour, incubated for 3 h in the dark with Act. D [10 µg · ml⁻¹] or water, and then re-irradiated with continous far-red light (after [170])

the effect of far-red light on anthocyanin synthesis should be modified in a predictable manner by specific inhibitors, e.g. by Actinomycin D, (= Act. D), a potent inhibitor of transcription, i.e. DNA-dependent RNA synthesis, or by Puromycin, an antibiotic which specifically inhibits protein synthesis (Fig. 70). This is indeed the case (Fig. 71). When Act. D [10 μg · ml⁻¹] is added to the solution surrounding the seedling before the onset of far-red light, no anthocyanin synthesis will occur. If, however, the antibiotic is added after the lag-phase, e.g. 6 h after the onset of light, anthocyanin synthesis will proceed in essentially the same way as in the controls but at a somewhat reduced rate (Fig. 72). These facts show that the total inhibition is not due to a non-specific effect of the Act.D. Obviously the seedlings can synthesize anthocyanin in the presence of the antibiotic if this synthesis is already under way at the moment when the substance is added to the system. We had to conclude, for a number of further reasons [144], that Act.D at a suitable dose will block the induction of anthocyanin synthesis by P_{fr} if applied before the onset of light but permits the continued synthesis of anthocyanin if the biosynthetic process has already started at the time the antibiotic is added.

In the case of Puromycin the effect on phytochrome-mediated anthocyanin synthesis is independent of the time of application (Fig. 73). If, for instance, Puromycin is applied

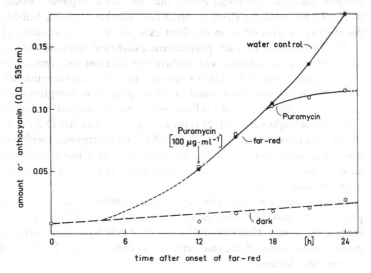

Fig. 73. Time-courses of anthocyanin accumulation in the mustard seedling under the influence of continuous far-red light with and without Puromycin [100 μg · ml⁻¹] in the solution surrounding the seedling (after [176])

after 12 h of far-red irradiation, it is observed that anthocyanin synthesis continues at an unaltered rate for several hours, but it virtually stops after about 5 h even in the case of continued far-red irradiation. This fact probably means that anthocyanin synthesis depends on continued undisturbed protein synthesis. One may speculate that at least one of the enzymes which are involved in anthocyanin synthesis decays below a critical level within 5 hours if protein synthesis is almost completely inhibited by Puromycin.

b) Enlargement of cotyledons. If an induction of transcription is involved in a "positive" response mediated by phytochrome, one might expect that Act.D would inhibit the response if applied before the onset of far-red light. This is clearly the case in far-red-mediated

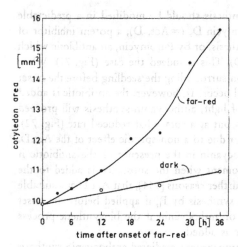

Fig. 74. The influence of continuous far-red light on the enlargement of mustard cotyledons. Onset of light: 36 h after sowing (after [163])

anthocyanin synthesis. However, if the same response does occur at a low rate even in complete darkness, one might expect that the "dark response" would not be inhibited by Act.D, provided the RNA involved does not have a short half-life. Enlargement of the mustard cotyledons is an excellent example of such a situation [175].

Fig. 74 illustrates the basic phenomena: a slight enlargement of the cotyledons in complete darkness, and a strong and qualitatively different enlargement under the influence of continuous far-red light. Table 4 presents data which indicate the Act. D [10 μg · ml⁻¹] will not influence the enlargement of the cotyledons in complete darkness, whereas the antibiotic cancels the far-red-mediated enlargement completely if applied at the onset of light. As one might expect, there is only a slight effect of Act.D if the antibiotic is applied 6 h after the onset of light. It was found [291] that the treatment with Act. D [10 μg · ml⁻¹] strongly reduces the increase of RNA even in darkness without interfering with the enlargement of the cotyledons under these conditions. It has been concluded that total RNA is not limiting for growth.

We have described only a few of the many inhibitor data which support the idea that phytochrome may sometimes exert its action through differential enzyme induction and enzyme repression. However, full support of this hypothesis can only be obtained by direct measurements of enzyme synthesis as controlled by P_{fr}. This will be the topic of the next few lectures.

Table 4. Enlargement of cotyledons of the mustard seedling under the influence of Act. D [10 μg · ml⁻¹] and continuous far-red light. Irradiation starts at 36 h after sowing; it ends 24 h later (after [175])

Treatment	Area of Cotyledon [mm²]
control₁ (36 h)	11.89 ± 0.18
control₂ (dark control, 60 h)	13.57 ± 0.17
far-red (no Act. D)	16.33 ± 0.22
far-red + Act. D (0 h)[a]	13.72 ± 0.41
far-red + Act. D (6 h)	15.42 ± 0.35
darkness + Act. D (0 h)	13.67 ± 0.33
darkness + Act. D (6 h)	13.37 ± 0.28

[a] time of application of Act. D

Let us close the present lecture by describing briefly a "pioneer" experiment which has contributed to open the way towards enzyme studies. If the function of P_{fr} is sometimes connected with a derepression of certain genes one might expect that RNA and protein synthesis would be activated if P_{fr} has been formed in a system – like the mustard seedling – which shows predominantly "positive" photoresponses. Table 5 indeed shows that both the RNA and the protein contents of a whole mustard seedling increase under the influence of $P_{fr\ (ground\ state)}$.

Table 5. Action of phytochrome on the RNA and protein levels of whole mustard seedlings, *Sinapis alba* L. (after [289])

	Program of Irradiation
"red"	36 h dark + 6× (5 min red + 355 min dark)
"far-red"	36 h dark + 6× (5 min far-red + 355 min dark)
"reversion"	36 h dark + 6× (5 min red + 5 min far-red + 350 min dark)
"control"	72 h dark

	% Increase of RNA and Protein Contents above Control		
	"red"	"far-red"	"reversion"
RNA	19.5	10.0	7.3
protein	10.1	3.4	4.4

Selected Further Reading

MOHR, H.: Differential gene activation as a mode of action of phytochrome 730. Photochem. Photobiol. 5, 469 (1966).

MOHR, H.: Untersuchungen zur phytochrominduzierten Photomorphogenese des Senfkeimlings (*Sinapis alba* L.). Z. Pflanzenphysiol. 54, 63 (1966).

MOHR, H., SITTE, P.: Molekulare Grundlagen der Entwicklung. München: BLV 1971.

WAGNER, E., MOHR, H.: "Primäre" und "sekundäre" Differenzierung im Zusammenhang mit der Photomorphogenese von Keimpflanzen (*Sinapis alba* L.). Planta 71, 204 (1966).

Phytochrome-mediated Enzyme Induction

Let us return to our favorite experimental material, the mustard seedling (cf. Fig. 1). In general, control by P_{fr} of enzyme synthesis[1] has been traced in the attached cotyledons of this seedling. In the course of the 5th lecture the hypothesis was advanced that P_{fr} in one and the same tissue or even cell can rapidly (i.e. in a matter of minutes, not hours) mediate enzyme induction as well as enzyme repression (Fig. 75). Further, we had to postulate enzymes whose synthesis or decay is not affected by P_{fr}. This 6th lecture will be concerned with the problem of verifying this model using the mustard seedling as our experimental material.

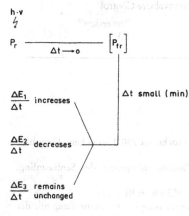

Fig. 75. This scheme may illustrate the hypothesis that P_{fr} in one and the same organ or tissue can rapidly mediate enzyme induction (above) as well as enzyme repression (middle). The symbol [$\frac{\Delta E_3}{\Delta t}$ remains unchanged] designates the postulate that some enzymes will not respond to P_{fr} (after [166])

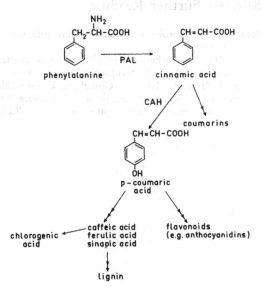

Fig. 76. This scheme may emphasize the significance of the enzyme phenylalanine ammonia-lyase (PAL) which can be regarded as the key enzyme of the phenylpropane anabolism (after [166])

1 While the use of the term "synthesis" seems to be justified by the results of density labelling (cf. Fig. 78) and of the usual inhibitor experiments (cf. Table 6) it must be realized that there is no rigorous proof that under all circumstances an increase of enzyme activity is paralleled by a corresponding increase in the number of enzyme molecules. Therefore the term "synthesis" (or "apparent synthesis") is used in the present text for convenience to denote any increase in enzyme activity and is not necessarily meant to imply true *de novo* synthesis.

A first example of phytochrome-mediated enzyme synthesis in the mustard seedling is L-phenylalanine ammonia-lyase, abbreviated PAL. This enzyme catalyzes the formation of trans-cinnamic acid from phenylalanine (Fig. 76). Obviously it is a key enzyme of flavonoid biogenesis and in this function it can be regarded as one of the most important enzymes of the secondary plant metabolism. Since synthesis of anthocyanin in the mustard cotyledons can be induced by P_{fr}, one could assume that P_{fr} induces PAL synthesis. This is indeed the case (Fig. 77). In the dark-grown mustard cotyledons PAL activity can hardly be detected. However, the enzyme can rapidly be induced by P_{fr}. The usual experiments with the usual inhibitors of RNA and protein synthesis (cf. Fig. 70) indicate that the increase of enzyme activity is due to an increase in the number of enzyme molecules. This conclusion

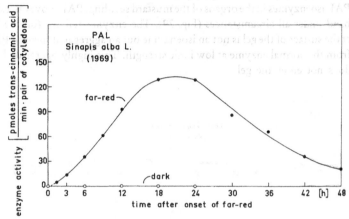

Fig. 77. The influence of continuous far-red light on the basic kinetics of the enzyme PAL in the cotyledons of the mustard seedling. Onset of far-red light: 36 h after sowing. In the dark-grown mustard cotyledons PAL activity cannot be detected by the assay which was used in this investigation (after [49])

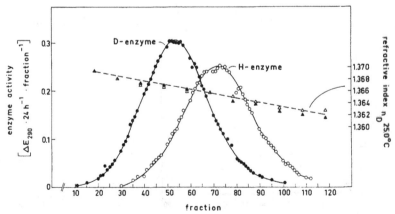

Fig. 78. This type of information supports the conclusion that the increase in PAL activity induced by continuous far-red light (or light pulses, cf. Fig. 81) must be attributed to a *de novo* synthesis of enzyme molecules. When the mustard seedlings are supplied with D_2O and irradiated with far-red light they form an enzyme, the density of which is considerably higher than the density of the PAL which is formed in seedlings which contain only H_2O. For details of the experimental method check the original paper [263]

has recently been confirmed by SCHOPFER and HOCK [263] using the *in vivo* density labelling technique with deuterium oxide. The experiments show unambiguously that a true *de novo* synthesis of PAL occurs in the cotyledons of the mustard seedling. Fig. 78 shows an equilibrium distribution of PAL in the CsCl-density gradient whereby the enzyme from seedlings in normal water, H-Enzyme, is shown together with the enzyme from seedlings which were supplied with deuterium oxide and illuminated with continuous far-red light. This enzyme species is designated D-Enzyme.

Another problem which must be dealt with before an interpretation of the PAL kinetics can be advanced is whether or not isoenzymes of PAL exist. Using improved methods of disc electrophoresis on polyacrylamide gel PETER SCHOPFER found no indications of PAL isoenzymes in the organs of the mustard seedling. PAL moves as a single, homogeneous band under all circumstances (Fig. 79). The enzyme activity (on the left) which remains at the surface of the gel is not an isoenzyme but an aggregated form of PAL which originates from the normal enzyme at low ionic strength. This highly-aggregated form of the enzyme does not enter the gel.

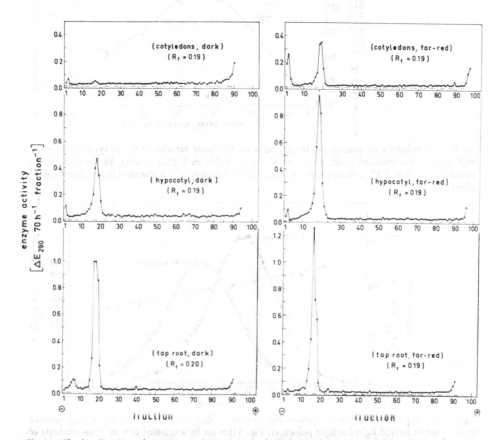

Fig. 79. The localization of PAL on polyacrylamide disc electrophoresis columns indicates that the enzyme synthesized under the influence of phytochrome is electrophoretically indistinguishable from the enzyme present in the organs of the dark-grown mustard seedling. The hypocotyl enzyme was concentrated 3×. For experimental details check the original paper [260]

Let us return to the question of how the far-red-mediated kinetics of PAL in the mustard cotyledons can be understood. The experimental and theoretical analysis of far-red → dark kinetics (cf. Fig. 164) has contributed the most towards answering this question. The basis of these experiments (as far as the phytochrome is concerned) is briefly the following: At the moment when the far-red light is turned off, the physiological effectiveness of a given P_{fr}-pool drops instantaneously to a low level. This statement is based on the fact that the physiological effectiveness of a stationary P_{fr}-pool (established by continuous far-red light) is a nearly logarithmic function of the quantum flux density over a wide range. In the 4th lecture we attributed the irradiance dependency to the action of P_{fr}^*, which is some excited species of P_{fr}. Let us postpone an analysis of these kinetics, which are measured after the far-red light is turned off, till the 16th lecture; at present I only want to summarize the principal results in Fig. 80. We have good reason to believe that the far-red-mediated PAL kinetics in the mustard cotyledons can be fully explained by three factors: induction of PAL synthesis by P_{fr} (whereby P_{fr} is continuously required); inactivation of PAL by

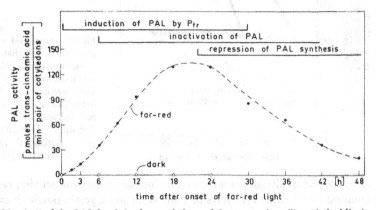

Fig. 80. The "basic kinetics" of the PAL levels in the cotyledons of the mustard seedling (dashed line) can be explained by an interaction of the following 3 factors: enzyme induction, enzyme degradation (inactivation), repression of PAL synthesis. The beginning and (in the case of the first factor) the end of the period of involvement of the factors are indicated by the vertical bars (after [49])

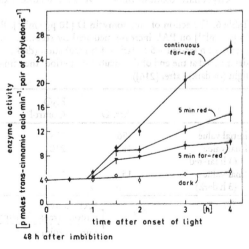

Fig. 81. Increase of PAL under continuous far-red light, or following a brief irradiation (5 min) with red or far-red light at time zero (= 48 h after sowing). The lag-phase is the same with all treatments. With the very sensitive assay used in these experiments a dark level of PAL can be detected even in the cotyledons (after [264])

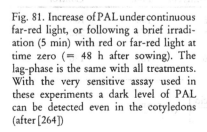

an inactivating principle; repression of PAL synthesis (i.e. for some reason or other the system no longer responds to P_{fr} with the formation of PAL).

I would like to say just a few words in this context about the problem of the initial, or primary, lag-phase. At 36 h or 48 h after sowing there is always a significant lag-phase, on the order of an hour, before PAL synthesis becomes measurable after the onset of continuous far-red light or after a red or far-red light pulse (Fig. 81). If, however, a seedling which has been preirradiated with 12 h of far-red light is kept in darkness for 6 h and is then reirradiated with far-red light, no lag-phase for the action of the second irradiation can be measured. Enzyme increase is instantaneous and *linear* (Fig. 82). Since the action

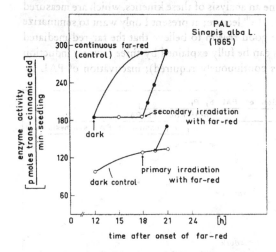

Fig. 82. Initial and secondary lag-phases of far-red-mediated increase of PAL in the mustard seedling. To determine the secondary lag-phase the seedlings were irradiated for 12 h with far-red light, placed in darkness for 6 h and re-irradiated with far-red light. Note that the whole mustard seedling served as a system of reference in these experiments (after [216])

of the second irradiation, as measured by increase of enzyme activity, can be completely inhibited by relatively low doses of Puromycin and Cycloheximide, it can be concluded that the reappearance of P_{fr} leads to *de novo* synthesis of enzyme protein (Table 6). Application of Act.D [10 µg · ml^{-3}] only partially inhibits the action of the second irradiation, as one would expect after the experience with anthocyanin synthesis (cf. Fig. 71, 72).

Glycollate oxidase may serve as a second example of an enzyme which is inducible

Table 6. The action of Actinomycin D [10 µg · ml^{-1}], Puromycin [100 µg · ml^{-1}] and Cycloheximide [5 µg · ml^{-1}] on PAL increase mediated by secondary irradiation with continuous far-red light. Program: 12 h far-red – 5 h dark – 1 h incubation (dark) – 3 h far-red (or dark). The "initial value" was determined at the end of the incubation period; the "final value" at the end of the 3 h period of far-red light (or dark) (after [216])

	Act. D	H$_2$O Control	Puromycin	Cyclohexi-mide	H$_2$O Control
initial value	156	156	186	186	186
final value (3 h far-red)	183	210	195	174	240
final value (3 h dark)	153	156	186	186	186

$$\left[\frac{\text{PAL Activity} \quad \text{p moles trans-cinnamic acid}}{\text{min · seedling}} \right] \quad (\text{cf. }[49])$$

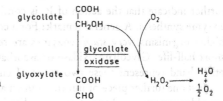

Fig. 83. The reaction catalyzed by the flavoprotein
glycollate oxidase

by P_{fr} in the cotyledons of the mustard seedling. Glycollate oxidase is a flavoprotein which
catalyzes the oxidation of glycollate to glyoxylate (Fig. 83). Glycollate oxidase has been
isolated from many plants. The enzyme seems to be localized in microbodies, in particular
in peroxisomes. Fig. 84 shows that glycollate oxidase (GO) can indeed be induced by
P_{fr} in the cotyledons of the mustard seedling. However, to some extent the enzyme can
even be synthesized in the dark. P_{fr} increases the rate of accumulation and the level of
enzyme which is eventually reached.

How can the kinetics of glycollate oxidase under the influence of continuous far-red
light be explained? Again the information of far-red → dark kinetics contributes to the
right answer. If the far-red light is turned off after 21 h, the increase of the enzyme gradually
stops and a constant level is established. If the far-red light is turned off after 48 h of
far-red light, no influence on the enzyme level can be detected. The simplest explanation
is that glycollate oxidase does not show any decay (or inactivation) during the experimental
period. The final enzyme level is reached when enzyme synthesis stops. The factors of
repression, i.e. those factors which are responsible for the termination of glycollate oxidase
synthesis under continuous far-red light, have no relation to phytochrome. These factors
are part of the process of "primary differentiation".

The behavior of glycollate oxidase seems to be representative of the majority of the
enzymes which can be induced by P_{fr}. I would like to mention briefly two further examples
which are very similar to glycollate oxidase, namely ascorbate oxidase and glyceraldehyde-
phosphate dehydrogenase. The terminal oxidase ascorbate oxidase can be induced by P_{fr}
in the cotyledons of the mustard seedling (Fig. 85). The far-red→ dark kinetics indicate
that enzyme inactivation does not significantly come into play in the cotyledons. They

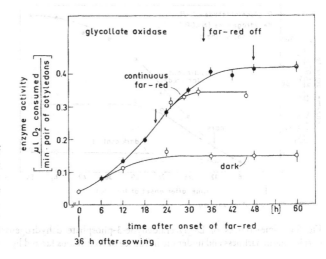

Fig. 84. Time-courses of
glycollate oxidase in the
mustard cotyledons in darkness and under the influence
of continuous far-red light.
Two far-red → dark kine-
tics (dark ↓) are indicated as
well (after [204])

further indicate that the action of P_{fr} is continuously required to maintain a high rate of enzyme synthesis. A general remark: For a consistent theory of regulation in multicellular higher organisms, very *stable* enzymes are required as well as enzymes with a relatively short half-life. Glycollate oxidase or ascorbate oxidase on the one hand and PAL on the other hand represent these two types of enzymes.

Just one further piece of information on ascorbate oxidase (cf. Fig. 85). If the seedlings are older enzyme synthesis starts immediately after the onset of far-red light, i.e. without an appreciable initial lag-phase. We realize, and this is again a general statement, that P_{fr} can *rapidly* induce enzyme synthesis if the seedling is in the proper state of development.

Enzymes which are localized in the chloroplast compartment can also be induced by P_{fr}. Without going into any detail at the moment, let me describe a single representative example, NADP-dependent glyceraldehyde-3-phosphate dehydrogenase (Fig. 86). The far-red → dark kinetics indicate that the enzyme is stable. They also indicate that P_{fr} is conti-

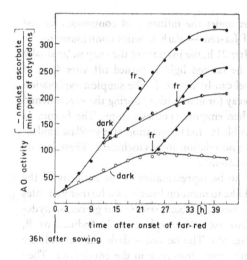

Fig. 85. Time-courses of ascorbate oxidase in the mustard cotyledons in darkness and under the influence of continuous far-red light. Note that the rate of enzyme increase drops immediately after the light as turned off and that there is no detectable initial lag-phase after the seedlings have reached a certain physiological age (after [52])

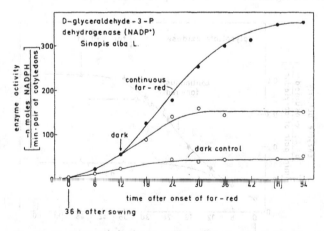

Fig. 86. Time-courses of D-glyceraldehyde-3-phosphate dehydrogenase (NADP$^+$) in the mustard cotyledons in darkness and under the influence of continuous far-red light (after [36])

nuously required to maintain enzyme synthesis. One should perhaps emphasize once more in this connection that even a long-term irradiation with our standard far-red light will not lead to a considerable chlorophyll formation (cf. Fig. 60). However, the yellowish plastids grow and reach the same size and dimensions as they do under white light.

It can be concluded, with respect to enzyme induction by P_{fr}, that even in the same tissue, namely cotyledons, the kinetics of induction of different enzymes can be very different. One reason for the differences is that some enzymes show a considerable decay rate while in other cases the rate of decay is very low, at least during the period which was investigated.

Another question we must answer in connection with our model (cf. Fig. 75) is the following: are there really enzymes in the mustard seedling whose temporal development, i.e. synthesis and decay, is not influenced by P_{fr} even if we induce dramatic photomorphogenic changes? The existence of such enzymes is a prerequisite for the model to explain photomorphogenesis on the basis of differential enzyme induction and repression. Isocitrate-lyase (a key enzyme of the glyoxylate cycle) may serve as an example of those enzymes that have synthesis time-courses in the cotyledons of the mustard seedling which are totally independent of P_{fr} (Fig. 87). Neither light pulses nor continuous far-red light (i.e. P_{fr}^*)

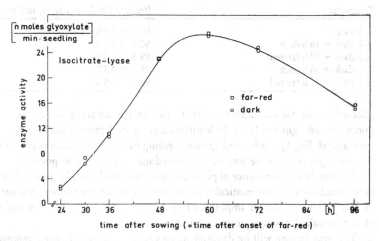

Fig. 87. Time-course of isocitrate-lyase in the mustard seedling in darkness and under the influence of continuous far-red light (after [127])

have a significant influence on isocitrate-lyase, although the enzyme does show a "grand period", i.e. a strong increase and a following decline of activity during the period of our experimentation. If the onset of far-red light is only 36 h after sowing, the same result is obtained. In any case we could not detect even the slightest influence of far-red light on the grand period of the enzyme. As one might expect, synthesis of isocitrate-lyase is hardly affected by Act.D whereas the application of Puromycin totally inhibits the increase of enzyme activity [127].

Catalase serves as another example of an enzyme whose time-course is hardly influenced by continuous far-red light (Fig. 88). It might be worthwhile to stress the following point with respect to enzymes like catalase or isocitrate-lyase (Table 7). Both the temporal development and the relative distribution of these enzymes within the seedling are not significantly

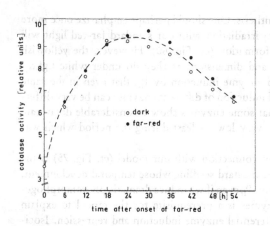

Fig. 88. Time-courses of catalase in the mustard seedling in darkness and under the influence of continuous far-red light. Onset of light: 36 h after sowing. It is not excluded that there is a slight inductive effect of the light during the second half of the kinetics (after [54])

Table 7. The relative amount of catalase activity present in the cotyledons of mustard seedlings grown in the dark or under continuous far-red light. The amount of catalase activity in the whole seedling is taken as 100 per cent (4 independent parallels) (after [54])

Treatment	Per cent Catalase Activity in Cotyledons
36 h dark	$88.0 \pm 0.9\%$
36 h dark + 18 h dark	$90.0 \pm 0.5\%$
36 h dark + 18 h far-red	$89.3 \pm 0.5\%$
36 h dark + 36 h dark	$88.2 \pm 0.5\%$
36 h dark + 36 h far-red	$90.6 \pm 0.5\%$

affected by the far-red light treatment. This fact is amazing insofar as a seedling grown under far-red light for 18 or 36 h differs a great deal from a dark-grown seedling of the same age (cf. Fig. 1): the far-red-grown seedling has large cotyledons and a short hypocotyl; the dark-grown seedling has small cotyledons and a long hypocotyl.

Obviously the occurrence of phytochrome-mediated morphogenesis is a specific phenomenon and does not automatically affect every aspect of metabolism and integration within the plant. This is another important point to bear in mind when discussing the question of the so-called primary reaction of P_{fr}.

The next lecture will be devoted to an example of P_{fr}-mediated *repression* of enzyme synthesis and to the question of what happens at the level of RNA during P_{fr}-mediated enzyme induction and repression.

Suggested Further Reading

Karow, H., Mohr, H.: Aktivitätsänderungen der Isocitritase (EC 4. 1. 3. 1) während der Photomorphogenese beim Senfkeimling. Planta 72, 170 (1967).
Mohr, H.: Regulation der Enzymsynthese bei der höheren Pflanze. Naturw. Rdsch. 23, 187 (1970).
Mohr, H., Sitte, P.: Molekulare Grundlagen der Entwicklung. Munchen: BLV, 1971.
van Poucke, M., Barthe, F.: Induction of glycollate oxidase activity in mustard seedlings under the influence of continuous irradiation with red and far-red light. Planta 94, 308 (1970).

Enzyme Repression, Mediated by Phytochrome through a Threshold Mechanism

In Fig. 75 in the 6th lecture we advanced the concept that P_{fr} in one and the same tissue (and possibly cell) will rapidly induce enzyme synthesis as well as repress enzyme synthesis. We also had to postulate further enzymes whose synthesis or decay are not affected by P_{fr}.

In the course of the previous lecture we dealt with several examples of phytochrome-mediated enzyme induction and we discussed enzymes whose grand period is not significantly affected by the presence or absence of P_{fr}.

We now proceed to the description of a case where enzyme synthesis[1] in the cotyledons of the mustard seedling is arrested by P_{fr}. In this case only $P_{fr\ (ground\ state)}$ seems to be involved in the response. There are no experimental indications that $P_{fr}{}^*$ is also involved. This fact has made our investigations considerably easier.

Lipoxygenase (or lipoxidase as it was formerly called) is a plant enzyme that catalyzes the oxidation of unsaturated fatty acids containing a methylene-interrupted multiple-unsaturated system in which the double bonds are all cis, such as linoleic, linolenic and arachidonic acids, to the conjugated cis-trans, hydroperoxide. The cotyledons of the dark-grown mustard seedling produce lipoxygenase at a relatively high rate. Fig. 89 shows that the

Fig. 89. The increase of lipoxygenase in the dark-grown mustard seedling is arrested by continuous standard far-red light. The far-red light maintains a P_{fr}/P_{total} ratio close to 0.025. The extrapolation indicates that lipoxygenase is a stable enzyme during the period of experimentation and that a lag-phase of the repression response is not detectable (after [193])

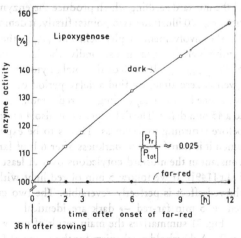

synthesis of lipoxygenase is arrested immediately after the onset of far-red light, i.e. after the formation of a relatively low but nearly stationary concentration of the effector molecule

---

1 The term "synthesis" is used in the present lecture for convenience to denote any increase of enzyme activity although *de novo* synthesis of lipoxygenase has not been demonstrated rigorously so far. However, the usual inhibitor experiments permit the conclusion [193] that an intact RNA- and protein synthesis is required for an increase of lipoxygenase activity.

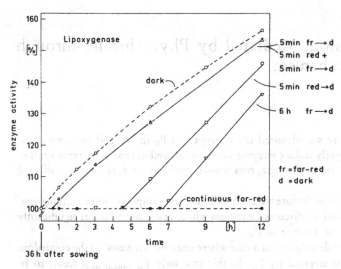

Fig. 90. The time-courses of lipoxygenase formation after a single light pulse at time zero with red, far-red, or red followed by far-red show that the operational criteria for the involvement of phytochrome (P_{fr} in the ground state) are fulfilled. The identical lag-period before resumption of enzyme synthesis in the cases 5 min far-red → dark and 6 h far-red → dark suggests that even in the case of long-time irradiation with far-red light only $P_{fr\ (ground\ state)}$ is involved (after [193])

P_{fr} in the seedling. The inhibition can be maintained for at least 12 h. Standard errors of measurement are between 0.3 and 1.5 per cent. This precision indicates that the cells of the mustard seedling which produce the enzyme form a fully-synchronized cell population. Fig. 90 illustrates two points: firstly, it demonstrates that the usual operational criteria for the involvement of phytochrome (P_{fr} in the ground state) in arresting lipoxygenase synthesis are fulfilled, and secondly, the data suggest a threshold mechanism for the action of $P_{fr\ (ground\ state)}$. If 80% of the total phytochrome is transferred to P_{fr} by 5 min of red light given at zero time, we find a delay period of 4.5 h in darkness before enzyme synthesis is resumed. A corresponding far-red irradiation which gives 2.5% P_{fr} at zero time leads to a 45 min delay. The delay period is also 45 min following six hours' exposure to far-red before returning to darkness. This is to be expected since 2.5% P_{fr} are left in a seedling when it is transferred to darkness after 6 h of far-red light. (Remember that P_{total} remains constant in the mustard cotyledons over at least 8 h after the onset of continuous far-red light [149]). If we follow 5 min of red light with 5 min of far-red light it turns out that the red effect is perfectly reversible. The two curves, 5 min far-red → dark and 5 min red + 5 min far-red → dark are identical.

Fig. 91 summarizes the main result of our work on control of lipoxygenase synthesis by P_{fr}. A threshold mechanism for the action of P_{fr} is postulated. Our data are in accordance with a half-life of 45 min for P_{fr} in darkness at 36 to 44 h after sowing, provided a level of 1.25 per cent P_{fr}, based on initial P_{total} at 36 h after sowing, is a threshold value for lipoxygenase synthesis.

If the amount of P_{fr} exceeds the threshold level lipoxygenase synthesis is fully and immediately arrested. If the amount of P_{fr} decreases below the threshold level, synthesis is immediately resumed at full speed. A threshold response like the one in Fig. 91 can be understood (in principle at least) on the assumption that repression of lipoxygenase

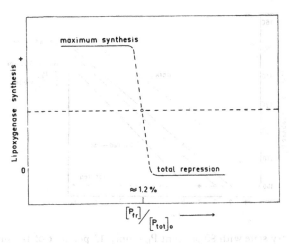

Fig. 91. A scheme to illustrate the concept of a threshold regulation of lipoxygenase synthesis by P_{fr} (ground state). $[P_{tot}]_0$, total phytochrome at time zero (36 h after sowing). This value is a constant. In the following the amount of P_{fr} is always expressed as per cent of $[P_{tot}]_0$. Expressed in this way, the threshold value of P_{fr} is in the neighborhood of 1.25 per cent

synthesis is a multiple-hit event which requires a cooperative model of the primary reaction of P_{fr}. In our context this means that a certain number ("critical number") of P_{fr} molecules must be available at a sensitive site (call it "control center") in order to perform repression of lipoxygenase synthesis. If the critical number is not fully reached, no effect of P_{fr} on lipoxygenase synthesis can be detected; however, if the critical number of P_{fr} molecules per control center *is* reached, the sensitive site is instantaneously and totally inactivated. As soon as the number of P_{fr} molecules per sensitive site decreases below the critical number, the "control center" is immediately and totally reactivated. It seems that the "critical number" of P_{fr} molecules which is required to inactivate a control center is high.

In order to understand the high precision of the threshold regulation of lipoxygenase synthesis, the following assumptions must be made:

1. Those cells of the mustard seedling which are involved in lipoxygenase production form a fully synchronized cell population (at least with respect to lipoxygenase synthesis).

2. Every cell of the population must have the same threshold (a).

3. Every cell of the population must have the same amount of phytochrome (b) (or the same fraction of total phytochrome if only a fraction is involved in regulation of lipoxygenase synthesis).

Postulates 2) and 3) could be combined to form the postulate that the value b/a must be constant in every cell of the population.

In addition it must be guaranteed that the individual seedlings of an experimental population behave in a highly synchronized manner (at least with respect to lipoxygenase). This requires an unusually precise control of every external factor, including the procedures of wetting and sowing the mustard seeds.

I shall mention only a few further experimental data which have contributed to the threshold concept. Remember in the following that the decay of P_{fr} in the mustard seedling follows a first order reaction with a half-life on the order of 45 min, at least between 36 h and 44 h after sowing.

If we establish 18% P_{fr} at time zero by irradiation of seedlings for 10 min with 705 nm radiation which fully establishes the P_{fr}/P_{tot} equilibrium, we can predict that after about 3 h (namely, four half-lives of P_{fr}) the threshold value will be reached. This is observed to be the case: enzyme synthesis is resumed at full speed after a lag of 3 h (Fig. 92).

If seedlings are irradiated for two hours with red light – which establishes a photostation-

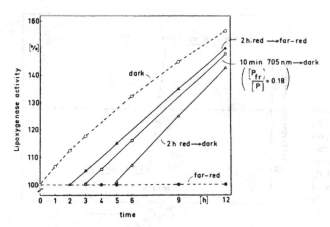

Fig. 92. Tests of the P_{fr} threshold requirement for lipoxygenase synthesis under several irradiation conditions. For details see text (after [193])

ary state with 80 per cent P_{fr} – only 13 per cent of the initial P_{total} (i.e. P_{total} at time zero) is still present in the seedling as P_{fr}. After transfer to darkness about 160 min in darkness is required for enzyme synthesis to resume. This is not too far from the prediction.

If the two-hour exposure to red light is followed by continuous far-red light, the situation with respect to P_{fr} is that far-red light establishes a photostationary state with about 2.5% P_{fr}. However, since the pretreatment with 2 h of red light has decreased the amount of total phytochrome by 84% (as compared with the amount of P_{total} at zero time) the value for P_{fr} divided by P_{total} at zero time is only 0.38 per cent, which is well below the threshold level of 1.25 per cent P_{fr}. Therefore, enzyme synthesis is immediately resumed under continuous far-red light (following a 2 hours' red light exposure) although the same light treatment (continuous far-red), without the prior exposure to red light, completely and continuously blocks enzyme synthesis.

This experiment has been very important to us, since it demonstrates that enzyme synthesis is resumed immediately (i.e. without a detectable lag) after the threshold level has been passed. Control of enzyme synthesis by P_{fr} is virtually instantaneous.

The next experiment presents evidence for a threshold based on synthesis of P_r (Fig. 93). The total phytochrome content of the mustard cotyledons can be decreased by red light. After one hour of red light, for example, only about 40 per cent of P_{tot} at zero time remains. If the cotyledons are then exposed to far-red light which transforms all but 2.5

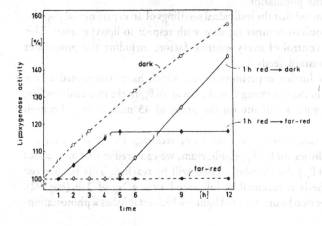

Fig. 93. A test of the threshold concept for control of lipoxygenase synthesis based on the synthesis of phytochrome (P_r). For details see text (after [193])

per cent of the P_{fr} back to P_r, the P_{fr} level is about 1 per cent of the initial P_{tot}. Enzyme synthesis is immediately resumed because the P_{fr} level is below the critical level of 1.25 per cent of P_{tot}. (Remember that the amount of P_{fr} is always expressed as a fraction of the total amount of phytochrome present at time zero, which is 36 h after sowing. Since this initial value is a constant, the term "per cent P_{fr}" represents the actual level of P_{fr}).

Phytochrome (i.e. P_r) synthesis under the continued exposure to far-red light raises P_{fr} by increasing P_{tot} until the photo-steady state is reached (cf. Fig. 28). The threshold level for stopping enzyme synthesis is thus attained after some period of time, in this case four and one-half hours from time zero. Obviously the newly-synthesized phytochrome is equal in properties and effectiveness to that initially present.

The control experiment, in which one hour of red light is followed by darkness, is also in accord with the threshold concept. The irradiation reduces P_{fr} to 32 per cent of P_{tot} at time zero. This amount decays to below 1.25 per cent P_{fr} in somewhat less than four hours. Thus, enzyme synthesis is resumed after a total of nearly 5 h from time zero.

Now let us proceed to a discussion of the result of our tests on reciprocity. The law of reciprocity is valid if the extent of a physiological effect (Δm) depends only on the total amount of quanta applied ($I \cdot t$), irrespective of intensity $= I$ (irradiance, quantum flux density). Expressed in symbols, this becomes $\Delta m = f (I \cdot t)$, whereby t is the time of irradiation.

Applied to our problem the situation is as follows: P_{fr} originates from P_r following a first order photochemical reaction. As long as only simple photochemistry is involved, reciprocity is expected to hold. Applied to mustard seedlings, this means that irrespective of irradiance, a given number of quanta per cm^2 is required to reach 1.25 per cent P_{fr}, the threshold level. The experimental results are indicated in Fig. 94. The standard far-red irradiance (100% irradiance) establishes 1.25 per cent P_{fr} so rapidly that the course of the reaction cannot be followed at the enzyme level. An immediate inhibition of enzyme synthesis is accordingly observed. With a 10 per cent irradiance, about 6 ± 3 min is required to stop enzyme synthesis, i.e., to reach the threshold level of P_{fr}. At 1.0 and 0.2 per cent of standard irradiance, the time is 56 ± 2 and 280 ± 4 min, respectively. Table 8 summarizes the main results. It is clear that the law of reciprocity is obeyed to the letter. The calculated

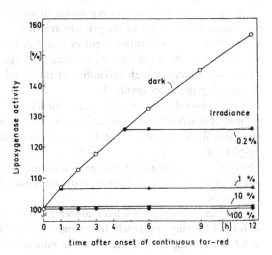

Fig. 94. A test for reciprocity in synthesis of lipoxygenase. The results indicate that irrespective of irradiance, a given number of quanta (Einsteins) per cm^2 is required to reach the threshold level of P_{fr} (after [193])

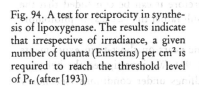

t-value for 100 per cent irradiance is 36 s, that is, at standard far-red irradiance it takes
about half a minute to achieve repression.

The data of Fig. 94 create problems which we cannot solve at the moment. Why does
reciprocity hold even under conditions (low irradiances) where the photochemical rate
constant is low compared with the rate constant of the P_{fr} decay? This problem is inherent

Table 8. The data of Fig. 94 are rearranged so as to show directly the validity of the law of reciprocity (after [193])

Irradiation Time t [s]	Irradiance[a] I[pE · cm^{-2} · s^{-1}]	I · t [pE · cm^{-2}]
36	2180	78 500
360	218	78 500
3 360	22	73 900
16 800	4.5	75 600

[a] light source: standard far-red light [162].

also in a number of earlier papers on phytochrome, like the one by WITHROW et al. [298]
on action spectra. The dose-response curves in this paper (determined with the highest
possible precision) show linearity in a semilogarithmic coordinate system down to very
low irradiances. They can only be understood if the P_{fr} decay is negligible. On the other
hand, there is no indication that P_{fr} decay does not occur at low P_{fr} concentrations [132,
219]. In what way are the P_{fr} molecules "counted" before they are subjected to the decay?
The answer to this (or similar) question(s) will possibly be required before we can understand the "primary action" of P_{fr} with respect to lipoxygenase control.

To summarize this part of the lecture:

Synthesis of the enzyme lipoxygenase in the cotyledons of the mustard seedling is
controlled by phytochrome through a threshold mechanism. The repression of enzyme
synthesis by P_{fr} is a very rapid process after the threshold level is exceeded. Conversely,
enzyme synthesis starts instantaneously and at full speed as soon as the P_{fr} level decreases
below the threshold level. Thus P_{fr} rapidly inhibits synthesis of an enzyme and functions
through an all-or-none control mechanism of high precision.

A threshold mechanism of regulation of enzyme synthesis very probably requires a
cooperative model as far as the primary reaction of $P_{fr\ (ground\ state)}$ is concerned. Furthermore a threshold mechanism requires a perfect synchronization of all those cells which
are involved in the response, i.e., produce lipoxygenase. Even if every cell responded in
a threshold manner, a high variability of the threshold would lead to a graded response
of the cell population involved.

Let us close this lecture with a few remarks on "photomorphogenesis as investigated
at the level of RNA". Under the influence of continuous far-red light the amount of RNA
per cell will increase in the cotyledons of the mustard seedling (Fig. 95). However, the
lag-phase of this response is at least 6 h and the profile of the RNA (as revealed by MAK-
Chromatography) does not change significantly. Therefore it can be concluded that the
increase of RNA per cell is a non-specific secondary result rather than a cause of photomorphogenesis.

After this experience we have concentrated over the years on short-term experiments
with the following apparently frustrating results (Fig. 96). We investigated the behavior
of RNA from the cotyledons of sterile mustard seedlings under conditions (secondary

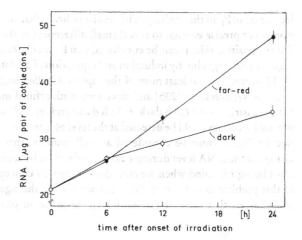

Fig. 95. Total RNA contents of the cotyledons of the mustard seedling growing in darkness or under the influence of continuous standard far-red light. Onset of light: 36 h after sowing (after [290])

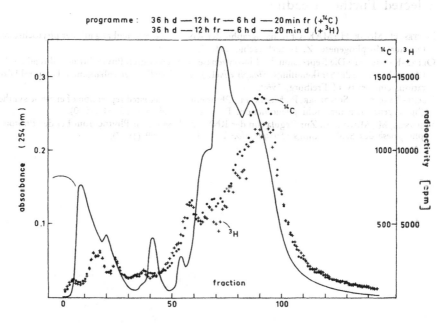

programme: 36 h d — 12 h fr — 6 h d — 20 min fr (+¹⁴C)
            36 h d — 12 h fr — 6 h d — 20 min d (+³H)

Fig. 96. Chromatography of double-labelled (^{14}C, ^{3}H) RNA from the cotyledons of sterile mustard seedlings on MAK-columns (MAK = methylated albumine on Kieselgur). Labelling was performed with ^{14}C-Uridine and ^{3}H-Uridine. Program: 36 h dark – 12 h far-red – 6 h dark – 20 min far-red (+ ^{14}C); 36 h dark – 12 h far-red – 6 h dark – 20 min dark (+ ^{3}H). The reciprocal experiment (i. e. dark-grown seedlings labelled with ^{14}C) leads to the same result. For experimental details check the original paper (after [48])

irradiation) where P_{fr} will rapidly mediate enzyme induction, e. g. phenylalanine ammonia-lyase induction (cf. Fig. 82). Using a double-labelling technique of RNA with ^{14}C-uridine and ^{3}H-uridine and separation of the RNA on MAK-columns we could not find any change in the RNA fractions within 20 min after the onset of light, i.e. within 20 min after the formation of P_{fr} in the seedling. The results indicate that the MAK method of RNA separation is not precise enough to reveal small differences at the level of rapidly-labelled RNA. On the positive side, it can be concluded that P_{fr} mediates morphogenesis not by dramatic regulatory changes but by induction and repression of a relatively small number of enzymes.

However, since at least most of the rapidly-labelled high molecular weight RNA seems to be rRNA precursor [235] and since even with refined methods no qualitative difference between dark-grown (36 h dark + 12 h dark) and far-red-grown (36 h dark + 12 h far-red) mustard seedlings could be detected at the level of transfer RNAs [47], a different interpretation of Fig. 96 must be considered as well, namely, that there is actually no qualitative change at the RNA level during enzyme induction by a secondary irradiation. This point must be kept in mind when we consider the current concepts of P_{fr} action. We shall return to this problem later (lecture 13) when we discuss the significance of the initial lag-phase and the fact that there is no secondary lag-phase in phytochrome-mediated responses.

Selected Further Reading

DITTES, H., MOHR, H.: MAK-Chromatographie der RNS im Zusammenhang mit der phytochromgesteuerten Morphogenese. Z. Naturforschg. *25b*, 708 (1970).

OELZE-KAROW, H.: Die Repression der Lipoxygenase-Synthese durch Phytochrom während der Photomorphogenese des Senfkeimlings (*Sinapis alba* L.), ein Schwellenwertsphänomen. Doctoral Dissertation, University of Freiburg, 1969.

OELZE-KAROW, H., SCHOPFER, P., MOHR, H.: Phytochrome-mediated repression of enzyme synthesis (lipoxygenase): a threshold phenomenon. Proc. nat. Acad. Sci. *65*, 51 (1970).

WEIDNER, M., MOHR, H.: Zur Regulation der RNS-Synthese durch Phytochrom bei der Photomorphogenese des Senfkeimlings (*Sinapis alba* L.). Planta *75*, 99 (1967).

Phytochrome-mediated Modulation of Metabolic Steady States and of Photonastic Movements

This 8th lecture will be concerned with the phenomenon of modulation. Let me first define this term. This definition is necessary since the term "modulation" has also been used recently in some fields of molecular biology in a completely different sense. In any theoretical treatment of development of multicellular higher systems we must discriminate between differentiation and modulation (Fig. 97). The term "modulation" is used to designate any change of a living system (or parts of it) which is fully reversed if the factor that has induced the change is removed. The term "differentiation" is used to designate those changes which remain even if the inducing factor is removed from the system. Although in some cases it depends on the system of reference (or on the point of view of the investigator) whether a particular process must be termed differentiation or modulation, in principle we must keep these two phenomena rigorously separate.

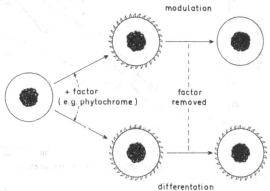

Fig. 97. Differentiation and modulation. In the case of "differentiation" the change (induced by the "factor") remains even if the inducing factor is removed. However, in the case of „modulation" the change (induced by the "factor") is fully reversed after removal of the inducing factor (modified after [292])

1. Ascorbic Acid Synthesis

Our first example is phytochrome-mediated control of ascorbic acid synthesis in the mustard seedling. Remember that the experimental material is the dark-grown mustard seedling growing under rigorously-controlled conditions at 25° C. The only variable factor is light.

The dark-grown mustard seedling between 36 and 60 h after sowing accumulates ascorbic acid at a low but nearly stationary rate (Fig. 98). P_{fr} (maintained in the seedling at an almost stationary concentration by continuous far-red light, the effective species being P_{fr}^{*}) increases the steady state rate of accumulation. If the far-red light is turned off, the rate decreases rapidly to the "dark" rate. If the far-red light is turned on again the rate rapidly increases, etc. It is evident that between 36 and 60 h after sowing the steady state rate of accumulation of ascorbic acid is reversibly modulated by P_{fr}. This type of control by P_{fr} does not depend on undisturbed RNA synthesis. It is, for example, not affected by Actinomycin D (Table 9). Under conditions where other "positive" photoresponses of the mustard seedling, such as anthocyanin synthesis or enzyme synthesis, are strongly inhibited by Actinomycin D, ascorbic acid synthesis is not significantly affected by the inhibitor. It can be concluded that regulation of gene activity is very probably not involved in photomodulation of ascorbic acid accumulation. You will see, if you look at the data more closely, that Act. D even seems to increase the rate of ascorbate synthesis somewhat. This effect is probably real. It is explained by the assumption that the inhibition of a number of biosynthetic pathways by Act. D will lead to an expansion of metabolic pools, including those which contribute to ascorbate synthesis.

In the case of control by P_{fr} of ascorbic acid production the operational criteria of a modulation are obviously fulfilled: the change introduced by the factor P_{fr} is fully reversed

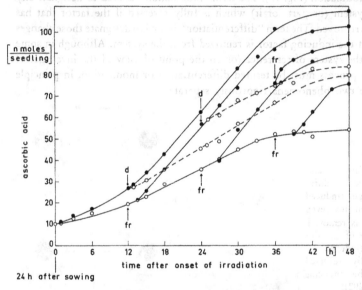

Fig. 98. Kinetics of ascorbic acid accumulation in the mustard seedling in the dark (–○–) and under the influence of standard far-red light (–●–). The arrows indicate the beginning (fr) or the end (d) of the irradiation period with far-red light (after [13])

Table 9. Content of ascorbic acid (AS) in the cotyledons of the mustard seedling, 24 h after the onset of far-red light. Concentration of Actinomycin D (Act. D) in the solution surrounding the seedlings and in the free diffusion space of the seedlings: [10 µg ml⁻¹] (adopted from [257])

Treatment	Amount of AS $\left[\dfrac{\text{n moles AS}}{\text{pair of cotyledons}}\right]$
dark	16.9
dark + Act. D (− 6 h)ᵃ	19.7
far-red	35.2
far-red + Act. D (− 6 h)ᵃ	35.7

ᵃ (− 6 h) indicates that the Act. D was applied 6 h before the onset of far-red light.

as soon as the factor is removed. But this statement is only true if we look at the metabolic chain which produces ascorbic acid. It is not true if we look at the ascorbic acid pool of a seedling or of an average cell. Seen from this aspect, far-red light causes a differentiation because the ascorbic acid contents remain at a higher level even if the far-red light is turned off. To repeat this statement in general terms: The modulation of a metabolic chain will lead to differentiation if the product of the chain is stable. However, if the product is not stable a metabolic modulation will not have consequences for the state of differentiation of a cell or tissue.

2. Carotenoid Synthesis

Before we proceed to the discussion of "modulation of organ movements" by P_{fr}, I would like to mention briefly a second example of a metabolic modulation in order to substantiate the assumption that this type of control occurs widely. Fig. 99 and Table 10 are concerned with carotenoid accumulation. The data show that the accumulation of carotenoids in the mustard seedling is controlled by P_{fr} ($P_{fr\ (ground\ state)}$ as well as $P_{fr}{}^{*}$). The time-courses of carotenoid accumulation in darkness and under the control of continuous far-red light

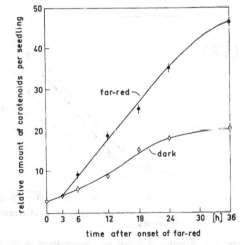

Fig. 99. Time-course of carotenoid accumulation in the mustard seedling in darkness and under the influence of continuous far-red light. Onset of light: 36 h after sowing. After an initial lag-phase of about 3 h, far-red light (i. e. $P_{fr}{}^{*}$) increases the rate of accumulation and prolongs the period of synthesis (after [267])

Table 10. Induction of carotenoid accumulation by red light and reversion of the induction by subsequent far-red light. Program: Irradiation (5 min each) with red, far-red or red followed by far-red light was performed 36 h and 42 h after sowing. Carotenoids were measured 60 h after sowing. It is clear that the operational criteria for the involvement of $P_{fr\ (ground\ state)}$ are fulfilled (after [267]). The same is true for ascorbate accumulation [254]

Program	Relative Amount of Carotenoids per Seedling [A]
dark control	0.219 ± 0.005
2×5 min red	0.273 ± 0.003
2×5 min far-red	0.221 ± 0.003
2× [5 min red + 5 min far-red]	0.222 ± 0.003

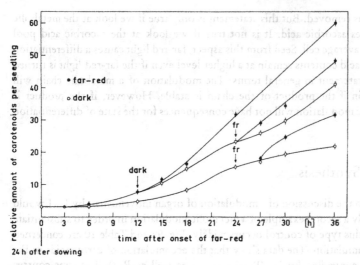

Fig. 100. Kinetics of accumulation of carotenoids in the mustard seedling in the dark (–o–) and under the influence of standard far-red light (–•–). The arrows indicate the end (d) or the beginning (fr) of the irradiation period with far-red light (after [267])

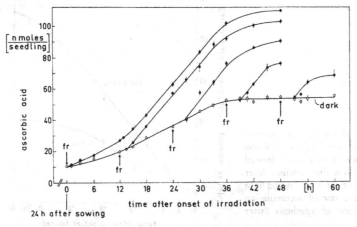

Fig. 101. Kinetics of ascorbate accumulation in the mustard seedling in the dark (–o–) and under the influence of standard far-red light (–•–). The time of onset of light was varied. The arrows indicate the onset of far-red light (after [13])

are considerably different. More detailed kinetic studies indicate (Fig. 100) that P_{fr} exerts a rapid and nearly reversible control over the rate of carotenoid accumulation. As is the case with ascorbic acid accumulation, accumulation of carotenoids is also insensitive towards Actinomycin D. Operationally, we have been using ascorbate and carotenoid synthesis as control parameters to make sure that there are non-specific effects of Act. D, at a concentration of [10 µg · ml^{-1}] in the solution surrounding and within the free diffusion space of the seedling.

One might expect that even in the case of a modulation, the possibility and the degree of a phytochrome effect would be determined by the primary state of differentiation (competence). This is indeed the case, for example, in P_{fr}-mediated ascorbic acid synthesis (Fig. 101). Up to about 30 h after sowing the mustard seedling hardly responds to P_{fr} as far as ascorbic acid synthesis is concerned. The full responsivity is only reached at 36 h after sowing. Accumulation of ascorbic acid tends to cease between 66 and 72 h after sowing irrespective of the onset of light and irrespective of the amount of ascorbic acid which has already been accumulated. 72 h after sowing the sensitivity of the ascorbic acid-producing system towards P_{fr} has nearly disappeared. Whereas the rate of accumulation in the linear part of the kinetics does not depend on physiological age, the onset and the termination of responsivity towards P_{fr} does depend on the physiological age of the seedling, i.e. on the pattern of primary differentiation, changing with time.

3. The Problem of the "Lag-phases"

At this point in the discussion I want briefly to deal once again with the problem of the lag-phase after the onset of light. This term, "lag-phase", is used to designate the duration of time which must pass after the onset of light before the response becomes measurable. There are two important aspects of the problem: Firstly, the length of the primary (or initial) lag-phase may differ from photoresponse to photoresponse. The term "primary lag-phase" is used to designate the lag-phase after the first onset of light with a previously dark-grown seedling. In the case of ascorbic acid synthesis the primary lag-phase *increases* as the seedling ages (Fig. 101), whereas in the case of anthocyanin induction the primary lag-phase is constant around 3 h over the whole range of experimentation (cf. Fig. 104). In the case of induction of phenylalanine ammonia-lyase the primary lag-phase gradually disappears as the seedling gets older (Fig. 102). This is the rule we have found so far with all enzyme inductions (cf. Fig. 85).

In no case could "secondary lag-phases" be detected. This term is used to designate the lag-phase under the treatment: light-dark period-reirradiation with the same light. One example (Fig. 103): if the light is turned off after 12 h of continuous far-red, the rate of ascorbic acid accumulation rapidly drops to the dark rate. If reirradiation is performed

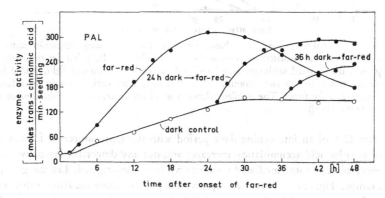

Fig. 102. Induction of phenylalanine ammonia-lyase (PAL) by standard far-red light. The duration of the initial lag-phase depends on the age of the mustard seedling. Time zero is 36 h after sowing. 72 h after sowing (36 h after time zero) an initial lag-phase can no longer be detected. Note that in this experiment the whole seedling was used as system of reference for PAL (after [166])

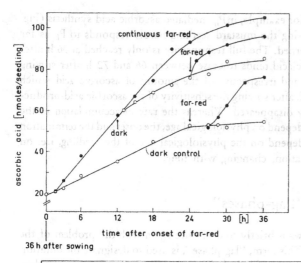

Fig. 103. Kinetics of ascorbic acid accumulation in the mustard seedling in the dark (–o–) and under the influence of standard far-red light (–•–). The arrows indicate the beginning (fr) or the end (d) of the irradiation period with far-red light. These data were selected to demonstrate that there is no secondary lag-phase even in the case of a modulation where the *initial* lag-phase is always considerable (cf. Fig. 101) (adapted from [13])

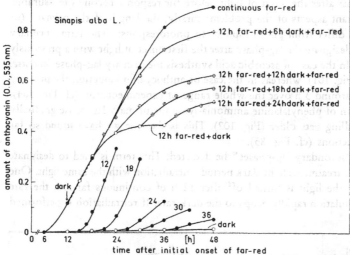

Fig. 104. Initial (or primary) lag-phases (lower part) and secondary lag-phases (upper part) of anthocyanin synthesis, mediated by continuous far-red light (i. e. P_{fr}^*). To determine the secondary lag-phases, the mustard seedlings were irradiated for 12 h with far-red light, placed in darkness, and reirradiated with the same continuous far-red light after dark intervals of different duration as indicated in the figure. The numbers (lower part of the figure) indicate the time of onset of far-red light (after [143])

after 12 h of an intervening dark period with the same source of far-red light, the rate of ascorbic acid accumulation increases without any detectable lag-phase. This is true for every phytochrome-mediated response we have investigated. Let me give just one more example (Fig. 104): The primary lag-phase in phytochrome-mediated anthocyanin synthesis is always around 3 h. If the far-red light is turned off after a primary irradiation period of 12 h and reirradiation is performed 12, 18 or 24 h later with the same far-red light, no lag-phase of the response can be detected. This is also true if we make use only of P_{fr} in the ground state by a brief irradiation with red light (Fig. 105).

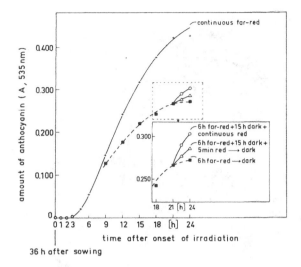

Fig. 105. An experiment to demonstrate that a primary irradiation (6 h standard far-red) eliminates the lag-phase for the action of continuous red or 5 min red (insert) even if a long dark period (15 h dark) is inserted between the primary and the secondary irradiation (after [145])

It can be concluded that the action of P_{fr} may always be fast, in differentiation as well as in modulation. Any lag-phases of the P_{fr}-mediated responses can always be overcome by a pretreatment with P_{fr}. Thus the actual mechanism can formally be described as

$$\frac{\Delta m}{\Delta t} = f \left(\frac{\Delta[P_{fr}]}{\Delta t}\right)$$

$$\Delta t_o \rightarrow 0$$

Expressed in words: a temporal change of P_{fr} is rapidly followed by a temporal change of the rate m of the response. The factor (or time) of delay, represented by Δt_o, is always small, even if we think in terms of rate constants of "molecular" responses, e.g. RNA or protein syntheses.

4. Modulations of Specialized Cells

Phytochrome is involved in a number of responses which are not related to morphogenesis (and possibly not even to differential protein synthesis). Rather, this type of phytochrome-mediated response must be classified under the term "fully-reversible modulations of highly-specialized cells". Unfortunately these phenomena have been greatly overestimated as far as their significance for the action mechanism of P_{fr} in *morphogenesis* is concerned (e.g. [76]).

One example, previously described in the 3rd lecture, is the phytochrome-mediated cloroplast movement in fully-differentiated algal cells; another example which will now be described briefly is the photonastic leaflet movement of the sensitive plant, *Mimosa pudica*. Both are rapid phenomena; under optimum conditions they can be fully expressed in 30 min.

a) Leaflet movement in *Mimosa pudica* (Fig. 106). Movements about the pulvini of the leaves, pinnae, and pinnules of the sensitive plant in response to touch and variations

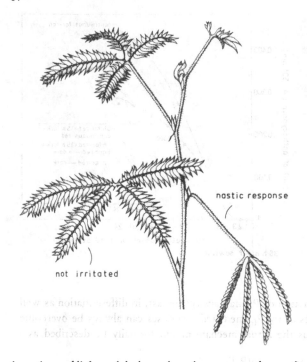

nastic response

not irritated

Fig. 106. A twig of the sensitive plant, *Mimosa pudica*, shows the nastic response of the leaves, pinnae and pinnules after irritation

in regimes of light and darkness have long attracted attention. However, only the Beltsville group has investigated this response thoroughly. We are only concerned here with the closing movements of pinnules on detached pinnae upon change from high-intensity fluorescent light to darkness (Fig. 107). This response depends upon the presence of phytochrome in the far-red-absorbing form, P_{fr}. In Fig. 107 detached pinnae are shown 30 min after transition from high-intensity fluorescent light to darkness. At the time of transition they were irradiated in succession for 2 min with red or far-red light to establish the corresponding photoequilibria (80 per cent *versus* about 3 per cent P_{fr}). The pinnae remained open if exposure to far-red was last, and closed if red radiation was last. Clearly, the closure movement of the pinnules is enhanced by P_{fr} and retarded if there is only a little P_{fr} present in the tissue. Since there is evidence that the tertiary pulvini, the sites of the response of the pinnules, act as the photoreceptors, and since it is known that potassium (K) fluxes are related to turgor changes in *Mimosa* motor cells, the *Mimosa* response has led to the conclusion that the P_{fr} in the pulvini acts on membrane permeability. This hypothesis has recently been supported by HILLMAN, GALSTON and associates [111, 119, 231, 232], working on nyctinasty in *Albizzia julibrissin*.

b) Leaflet movement in *Albizzia julibrissin* (Fig. 108). Various leguminous plants, including *Albizzia* species, have doubly compound leaves with paired leaflets that regularly move towards and away from each other in response to a variety of endogenous and exogenous stimuli. Leaflets usually open in the light and close when put in the dark, but their movement will persist with a circadian rhythm if plants are kept in constant-intensity light or in uninterrupted darkness. During certain parts of the diurnal cycle, nyctinastic closure is controlled by P_{fr}.

Leaflet orientation depends upon the relative turgidity of motor cells on dorsal and ventral sides of the pulvinule (= pulvinus of higher order = thickened organ at the base of every leaflet). GALSTON and associates [229, 230, 231] found that *Albizzia* leaflets close in the dark when potassium (K) moves into dorsal motor cells and out of ventral motor cells, and that red light (i. e. P_{fr}) is required for optimum K efflux from ventral cells and for optimum K movement into dorsal cells.

With *Albizzia* plants grown under a 16 h photoperiod, leaflets open rapidly upon transfer from dark to light, begin to close after 12 to 14 h, and usually close completely before darkness. They remain closed during the 8 h dark period, but will open without light if the usual dark period is extended 2 or 3 hours. During the diurnal cycle the K content of both dorsal and ventral motor cells changes about threefold. This has been interpreted as evidence for endogenously rhythmic changes in membrane permeability.

The interaction of the endogenous rhythm and P_{fr} in the *Albizzia* leaflet response can be summarized as follows: a high P_{fr} level is associated with a low K content in ventral motor cells, high K content in dorsal motor cells, and a small angle between leaflets (cf. Fig. 108). In other words, P_{fr} promotes closure of the leaflet pairs by optimizing K fluxes, but only within the range of response which is determined by the state of the endogenous rhythm. The action of P_{fr} does not depend on intact RNA synthesis (it is not sensitive towards Act.D).

The opening of *Albizzia* leaflets during appropriate parts of the diurnal cycle is promoted by light. In this response blue light is predominantly active. It is not a phytochrome-mediated response.

The control of leaflet movement by P_{fr} has been interpreted in a somewhat nebulous way as an effect of P_{fr} on "membrane permeability" (e.g. [76]). While this is possibly true for the motor cells of the pulvini (or pulvinules), positive evidence that differential changes in membrane permeability are universally connected with the primary action of phytochrome is lacking. Any extrapolation to the mode of action of P_{fr} in morphogenesis [76] is not justified.

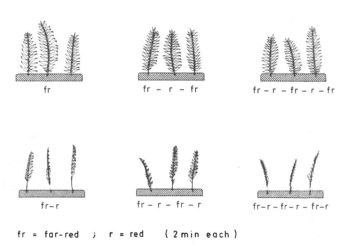

fr = far-red ; r = red   (2 min each)

Fig. 107. Pinnae of *Mimosa pudica*, 30 min after transition from high-intensity fluorescent light to darkness. At the time of transition they had been irradiated in succession for 2 min with red or far-red light to establish phytochrome predominantly in the P_{fr} or P_r form. The pinnae remained open if exposure to far-red was last (top row) and closed if red radiation was last (bottom row) (after [68])

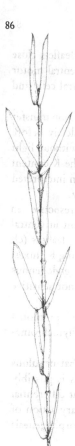

Fig. 108. Part of the leaf of *Albizzia* after transfer from light to darkness. The pulvini at the base of the second, fourth and sixth leaflets from the top were exposed to 2 min of far-red light immediately before the dark period. It is clear that far-red light has prevented the closure of only those leaflets whose pulvini were irradiated (adopted from [139])

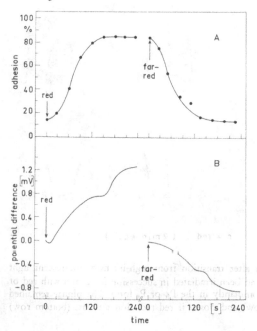

Fig. 109. The kinetics of (A) root tip adhesion to a negatively-charged glass surface ("Tanada effect") and (B) the development of electrical potentials in the root tip following irradiation with red or far-red light (after [166])

The so-called "Tanada effect" must be interpreted with the same caution. TANADA discovered [268] that tips of excised secondary roots of mung bean develop the ability to adhere to a negatively-charged glass surface when irradiated for 4 min with red light, and that this effect can be reversed by far-red light. Examination of the electrical potentials on the root tip showed [116] that red light caused the appearance of a positive charge on the surface of the root tip and that this electrical potential disappeared under the influence of far-red light (Fig. 109). Clearly, P_{fr} in some systems can lead to the development of bioelectric potentials. However, the physiological significance of these phenomena remains to be demonstrated. In any case all these phenomena are modulations, i.e. fully-reversible changes in specialized cells, not related to morphogenesis.

A final comment: The nebulous phrase "Phytochrome acts on a system such as membrane permeability" [76] must be replaced by the construction of solid models which can be subjected to an experimental test. The building of theoretical models could start from the assumption that the phytochrome molecule is a constituent of a membrane (e.g. plasmalemma), that it can act as a carrier, and that it possesses the properties of an allosteric protein. P_{fr} could be thought of as an active carrier while P_r could be thought of as inactive. If it is confirmed that protein synthesis (but not RNA synthesis) is required for P_{fr} to act on ion transport, the hypothesis could be advanced that P_{fr} rapidly acts on the level of translation by mediating the formation of new and specific carrier protein molecules, while P_r does not. These carrier protein molecules must have a very short half-life. In any case rigorous and quantitative models are required which can be checked experimentally. The "friendly persuasion" which is inherent in the term "action on membrane permeability" must be replaced by clear scientific statements which explicitly take into account the complexity of the situation (active transport, interaction with endogenous rhythm, specificity, etc.).

Selected Further Reading

BIENGER, I., SCHOPFER, P.: Die Photomodulation der Akkumulationsrate von Ascorbinsäure beim Senfkeimling (*Sinapis alba* L.) durch Phytochrom. Planta *93*, 152 (1970).

FONDEVILLE, J.C., BORTHWICK, H.A., HENDRICKS, S.B.: Leaflet movement of *Mimosa pudica* L. indicative of phytochrome action. Planta *69*, 359 (1966).

KOUKKARI, W.L., HILLMAN, W.S.: Pulvini as the photoreceptors in the phytochrome effect on nyctinasty in *Albizzia julibrissin*. Plant Physiol. *43*, 698 (1968).

MOHR, H., SITTE, P.: Molekulare Grundlagen der Entwicklung. München: BLV, 1971.

SATTER, R.L., GALSTON, A.W.: Phytochrome-controlled nyctinasty in *Albizzia julibrissin* III. Interactions between an endogenous rhythm and phytochrome in control of potassium flux and leaflet movement. Plant Physiol. *48*, 740 (1971).

SATTER, R.L., GALSTON, A.W.: Potassium flux: a common feature of *Albizzia* leaflet movement controlled by phytochrome or endogenous rhythm. Science *174*, 518 (1971).

SCHNARRENBERGER, C., MOHR, H.: Carotenoid synthesis in mustard seedlings as controlled by phytochrome and inhibitors. Planta *94*, 296 (1970).

Control of Longitudinal Growth by Phytochrome

1. General Remarks

The study of growth has been an important part of plant physiology especially since the term "growth" was narrowed to the meaning of "cellular lengthening" and auxin was generally regarded as being *the* growth-regulating factor. However, the molecular basis of growth regulation is still unsolved: the action mechanism of auxin is still under debate, as is the significance of extensin [142] and the existence of growth-limiting proteins [38]. There are a number of plant physiologists who feel that the importance of auxin has been greatly overestimated in the past and that the intense studies of auxin action in *excised* plant parts have possibly led to an unsatisfactory concept as far as the significance of auxin for the growth of an *intact* plant is concerned (cf. [76]). Indeed, there are very few instances in which the application of auxin to intact plants led to any increase in elongation at all. While the application of auxin to entire plants does produce effects on the orientation of leaves and sometimes curvatures of stems, an induction of significant alterations of the overall growth rate of stems and roots has not been observed. These facts have led some to the belief that auxin does not really control the rate of elongation under conditions of symmetrical straight growth in intact plants. Instead, some plant physiologists prefer to believe that auxin in the intact plant is mainly concerned with tropisms (cf. [76]).

I shall take up this point again in the lecture on phototropism. At the moment I would only like to make the assumption that there is no interaction as far as control of growth is concerned between light on the one hand, and auxin or gibberellin on the other hand.

2. Interaction or no Interaction between Phytochrome and Gibberellic Acid (GA₃) in Control of Hypocotyl Elongation in Mustard Seedlings [168]

a) *The theoretical background* [165, 259]. To analyze the simultaneous action of two factors on hypocotyl lengthening we shall use the theoretical model (Fig. 110) that if two factors act independently on a system A, the response of which – as measured by the change of *a steady state rate* – is designated by $\Delta A'$, there are only two possibilities for "no interaction". These are the following:

1) If the two factors act independently on the same causal sequence, "multiplicative calculation" (or "multiplicative behavior") of the two factors will be observed.

$$\text{Formula: } \Delta A'_{F_2} = a \cdot A'_{F_1} \text{ or } \Delta A'_{F_1, F_2} = a \cdot \Delta A'_{F_1}$$

Expressed in words: the *relative* change caused by a given dose of the second factor is always the same, irrespective of the extent of A' which is maintained by the other factor, *or:* the rate with two factors, applied simultaneously, is a defined fraction of the rate which

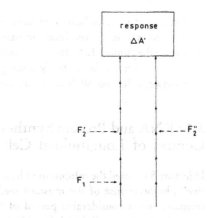

Fig. 110. This model reflects a situation in which two factors (F_1, F_2) simultaneously act on a system A. Only those changes of the system, responses $\Delta A'$, which can be conceived of as *changes of steady state rates* should be used in the analysis. It is assumed that one and the same $\Delta A'$ can be elicited through two totally separate reaction sequences. The model is explained in the text (modified after [259])

is obtained with one factor irrespective of the magnitude of the rate $\Delta A' F_1$. If a is > 1, the two factors act in the same direction, and if a < 1, the two factors act against each other.

2) The theoretical alternative is that the two factors act on different causal sequences which lead to changes of A, fully independently. An example would be that ascorbic acid is produced in 2 compartments of the cell independently and under the control of different factors. In terms of our model (Fig. 110): The factor F_1 controls the rate in compartment$_1$, the factor F''_2 controls the rate of ascorbic acid accumulation in compartment$_2$.

Under these circumstances, "numerically additive calculation" (or "numerically additive behavior") of the two factors will be observed.

$$\text{Formula: } \Delta A'_{F_1, F''_2} = \Delta A'_{F_1} \pm \Delta A'_{F''_2}$$

Expressed in words: A change of the rate A, $\Delta A'$, which is caused by the two factors F_1 and F_2 (if applied simultaneously) is identical with the sum of the changes caused by the two factors if applied separately. *To put this another way*, multiplicative as well as numerically additive behavior of the two factors indicate independence, i.e. no interaction. Any other experimental outcome means "interaction".

b) *The experimental data.* Fig. 111 summarizes some data concerning hypocotyl lengthening of the mustard seedling as controlled by GA_3 and red light simultaneously. The analysis strongly indicates that the total effect of the two factors on hypocotyl lengthening is composed in a numerically additive manner of the inhibiting light effect and the promotive GA_3 effect. You see that the dose-response curve for GA_3 is numerically the same, irrespective of the growth rate which is determined by light. The conclusion has been drawn [168] that P_{fr} and GA_3 do not interact as far as control of hypocotyl lengthening is concerned. This means that the control of axis growth by P_{fr} is certainly not mediated through GA_3.

The general conclusion (less justified) is that it is improbable that P_{fr} exerts its influence on symmetrical straight growth of the intact mustard seedling through the mediation of so-called growth hormones. However, since the intact seedling does not respond significantly to externally-applied auxin, this proposition cannot be checked experimentally with respect to auxin.

It has been found [258] that phytochrome and exogenously-applied "substrates" (ions, sucrose) act in a multiplicative manner, i.e. as independent factors, on hypocotyl lengthening. This is true for the intact mustard seedling as well as for the isolated hypocotyl (including the plumule!). Sucrose and ions (Knop's solution) strongly promote lengthening, which was found to be a steady state process under all experimental circumstances (at least between 36 and 60 h after sowing).

3. Is RNA and Protein Synthesis Related to Phytochrome-mediated Control of Longitudinal Cell Growth?

In lecture 5 we used the inhibition of hypocotyl lengthening by P_{fr} as prototype of a "negative" photoresponse of the mustard seedling. We recall that the growth rate is virtually constant over a considerable period of time in the dark as well as under the influence of continuous far-red light (cf. Fig. 65). Unfortunately the state of the phytochrome system in the hypocotyl (Fig. 112) is somewhat different from that in the cotyledons (cf. Fig. 27). Since the rate of synthesis of P_{fr} is much lower than in the cotyledons, the decay of P_{fr} dominates P_r synthesis even under standard far-red light. The consequence is that total phytochrome decreases slowly but steadily under continuous far-red light (Fig. 112). However, the deviation from a steady state is not too serious within the period of time used for experimentation. In every other respect the mustard hypocotyl is well-suited to the desired type of investigation. Firstly, longitudinal steady state growth of the organ is probably exclusively due to lengthening of the hypocotyl cells [80]. This means that

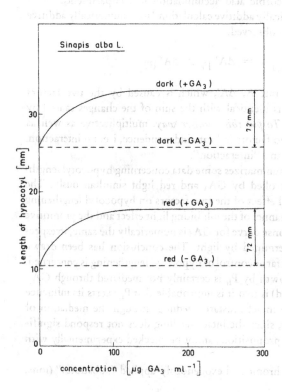

Fig. 111. An empirical example of "numerically additive behavior" of two factors. Hypocotyl lengthening in the mustard seedling was investigated under the control of light (continuous standard red light) and exogenously-applied gibberellic acid (GA₃). It is apparent that the dose-response curve for exogenously-applied GA₃ is the same with and without light. Note that GA₃ promotes lengthening whereas red light (via P_{fr}) inhibits lengthening of the hypocotyl. Hypocotyl length was measured 72 h after sowing (after [168])

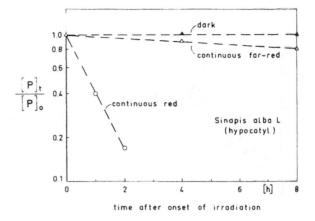

Fig. 112. Changes of total phytochrome [P] in darkness and under
continuous red or far-red light.
Material: Hypocotyl tissue of
mustard *(Sinapis alba)*. [P]$_0$, total
phytochrome 36 h after sowing
(time zero); [P]$_t$, total phytochrome at time t (time after onset
of light) (after [222])

hypocotyl lengthening represents lengthening of the average cell during the period of experimentation. Histological investigations show [80, 88] that cell division in the cortex of dark-
grown lettuce or mustard seedlings (standard conditions, 25° C) ceases between 55 and
60 h after sowing, whereas cell lengthening continues for several more days. Also, between
36 and 60 h after sowing the influence of cell divisions on cortex lengthening is nearly
negligible. In any case longitudinal growth of the cortex is predominantly due to cell lengthening. Secondly, since to do labelling experiments it is necessary to avoid translocation
from the cotyledons to the hypocotyl, the seedlings are used without cotyledons (Fig. 113).
The necessary operation is relatively simple and can be performed rapidly. However, there
is the risk that the removal of the cotyledons will modify the behavior of the seedling
remainder ("rest-seedling"). Fortunately, it is possible to run an experiment to check on
this problem. The results showed that the control of hypocotyl lengthening by continuous
standard far-red light is exactly the same with and without cotyledons (Table 11). This
means that phytochrome and the "factors of growth" which are supplied by the cotyledons
do not show any significant interaction. Rather, they act as independent factors in a multiplicative system.

About 15 min after the onset of continuous standard far-red light a new reduced steady

Table 11. The control of hypocotyl lengthening by far-red light is independent of the absolute growth
rate and independent of whether or not the cotyledons are attached to the axis system of mustard
seedlings (after [256])

Hypocotyl Lengthening	Dark	Far-red
growth rate [mm · h^{-1}]:		
intact seedlings	0.79	0.14
seedlings without cotyledons	0.39	0.07
inhibition by far-red [%]:		
intact seedlings	0	82
seedlings without cotyledons	0	81
inhibition due to the lack of		
cotyledons [%]	54	50

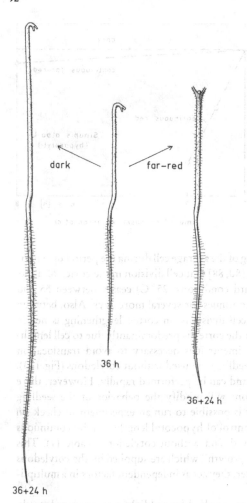

dark   far-red

36 h

36+24 h

36+24 h

Fig. 113. These drawings perhaps illustrate the usefulness of the axis system without cotyledons ("rest-seedling") for the causal analysis of hypocotyl lengthening in the mustard seedling (check Table 11). The total nitrogen of the axis system (hypocotyl plus radicle) does not change during the experimental period (after [172])

state growth rate is established (cf. Fig. 65). It differs from the growth rate in the dark by a factor of 8. In order to measure RNA and protein synthesis as a function of growth rate the seedlings (without cotyledons) are incubated with ^{14}C-uridine or ^{14}C-leucine up to 30 min after the onset of light. At the end of the period of incubation the surface and the free diffusion space of the hypocotyls are freed from label by rinsing briefly (5 min). This procedure – a brief rinsing at the end of incubation – eliminates more than a third of the label. Further rinsing (up to 30 min) has no further effect. This means that the label which remains in the hypocotyl after 5 min rinsing is inside the plasmalemma.

The main results which were obtained by the combined efforts of several investigators (K. Roth, W. Link, C. Holderied) are briefly the following:

a) If we follow the protein contents of the hypocotyl we find the amazing result (Fig. 114) that the total protein of the organ decreases steadily in spite of the fact that the organ grows at a constant rate. Further, there is no significant difference in protein contents between dark-grown and far-red-grown systems, in spite of the fact that the growth rates differ by a factor of 8 (cf. Fig. 65).

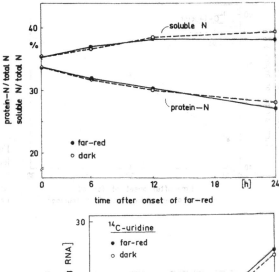

Fig. 114. Changes of the contents of protein and soluble N-containing compounds in the mustard hypocotyl in darkness and under the influence of continuous far-red light in a system without cotyledons (cf. Fig. 113). Total N in the whole rest-seedling (hypocotyl plus taproot) is a constant (after [120])

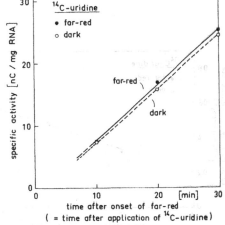

Fig. 115. The incorporation of ^{14}C-uridine into RNA of the hypocotyl of the rest-seedling (cf. Fig. 113) in the dark and under continuous far-red light (after [222])

b) RNA synthesis (as measured by incorporation of ^{14}C-uridine into RNA) is linear over at least 30 min after the onset of the experiment (Fig. 115). There is no significant difference between dark-grown and far-red-grown mustard seedlings, in spite of the growth rate differing by a factor of 8.

c) Likewise, protein synthesis (as measured by incorporation of ^{14}C-leucine into protein) remains unchanged for at least 30 min after the onset of continuous far-red light (Fig. 116).

d) The total uptake (corrected for free diffusion space by rinsing briefly) of the ^{14}C-precursors is not influenced by far-red light. This is true for uridine (Fig. 117) as well as for leucine (Fig. 118).

Unfortunately the internal cold pools of uridine or leucine are not known and can probably not be determined unambiguously [197]. However, it is most improbable that these pools have increased 8-fold within a few minutes after the onset of far-red light. Rather, the data indicate that only a very small fraction, if any, of total RNA and protein synthesis in a hypocotyl cell is directly connected with growth. The data do not support the view that changes of permeability or changes in the rate of active transport are early consequences of phytochrome action in general.

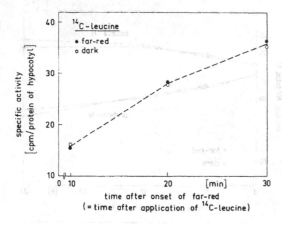

Fig. 116. The incorporation of ¹⁴C-leucine into protein of the hypocotyl of the rest-seedling in the dark and under continuous far-red light (after [222])

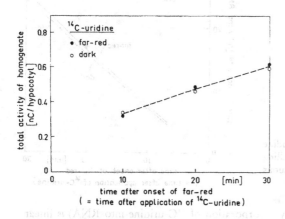

Fig. 117. Total uptake of ¹⁴C-uridine into the hypocotyl of the rest-seedling in the dark and under continuous far-red light. Remember that the free diffusion space was emptied by 5 min rinsing (after [222])

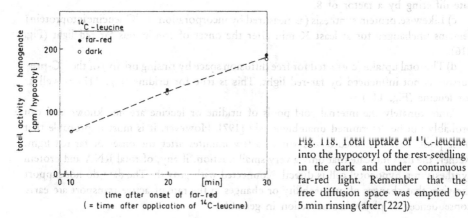

Fig. 118. Total uptake of ¹¹C-leucine into the hypocotyl of the rest-seedling in the dark and under continuous far-red light. Remember that the free diffusion space was emptied by 5 min rinsing (after [222])

4. Is Carbohydrate Metabolism Related to Control of Longitudinal Cell Growth?

A.M. STEINER [245, 246] has investigated carbohydrate metabolism intensely in connection with phytochrome-mediated inhibition of hypocotyl lengthening. Since none of the observed changes could be shown to be quantitatively or even qualitatively connected with phytochrome action, let me mention a few selected results only, which illustrate the principle and might be of general interest:

a) The amounts of free sucrose and of reducing sugars in the hypocotyl change dramatically under the influence of far-red light, but much later (hours later) than the growth rate (Fig. 119). Clearly, these effects are late consequences rather than causes of the reduced growth rate. These figures are an impressive demonstration of the statement that only kinetics, i. e. time-courses of a response, count in physiology. If end-point determinations only were available in this case, e.g. after 12 h, an appropriate interpretation of the data would be impossible, or, if advanced, the interpretation would be totally misleading.

b) The amounts of free glucose and fructose in the hypocotyl of the mustard seedling also change as a consequence of the action of far-red light, but once again these effects are a consequence rather than a "cause" of the reduced growth rate (Fig. 120).

c) The quantity of the main cell wall constituents of carbohydrate nature, as represented by the amounts of the monomeric sugars and uronic acids, responds only very late and only slightly to the growth-inhibiting action of continuous far-red light (Fig. 121). This has proved to be a very important result. It means that the synthesis of the cell wall constituents is much less inhibited than is cellular lengthening. There is no simple correlation between cellular growth and synthesis of cell wall constituents.

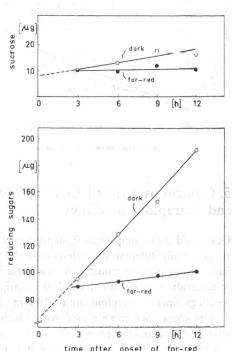

Fig. 119. The amounts of soluble sucrose and reducing sugars in the mustard hypocotyl in the dark and under continuous far-red light. Onset of light: 36 h after sowing (after [245])

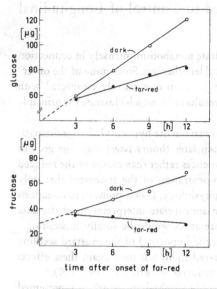

Fig. 120. The amounts of soluble glucose and fructose in the mustard hypocotyl in the dark and under continuous far-red light. Onset of light: 36 h after sowing (after [245])

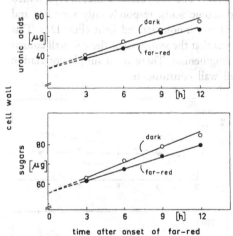

Fig. 121. The amounts of cell wall carbohydrates of the mustard hypocotyl in the dark and under continuous far-red light. Onset of light: 36 h after sowing (after [245])

5. Control by Far-red Light of Hypocotyl Lengthening in Diploid and Tetraploid Seedlings

One would like to compare the P_{fr}-dependent photoresponses of seedling populations which are genetically different, the difference of the genetical constitution being exactly known. Such a desire is understandable in view of the immense contribution of microbial genetics to molecular biology. We decided to investigate the responses of diploid and tetraploid seedlings. Since no tetraploids are available in white-seeded mustard (*Sinapis alba*), seedlings of *Oenothera hookeri* were used with which homogeneous populations of diploid and tetraploid seed could be grown. The main results (Fig. 122) were as follows: As one might expect, hypocotyl lengthening (among other photoresponses) is controlled by continuous far-red light (probably P_{fr}^*) in the case of *Oenothera* seedlings too. The further results, however, were not quite satisfactory. In fact it turned out that in darkness the growth curves of the hypocotyls are almost the same in diploid and tetraploid seedlings. Hypocotyl lengthening in the diploids is obviously not limited by the number of allelic genes per cell. Unfortunately, under the influence of far-red light the kinetics of hypocotyl lengthening are again identical in diploids and tetraploids.

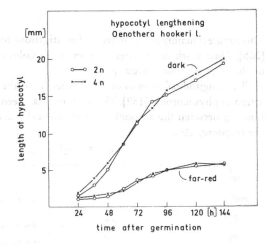

Fig. 122. Hypocotyl lengthening in diploid (2n) and tetraploid (4n) seedlings of *Oenothera hookeri* in darkness and under the influence of continuous far-red light (after [20])

6. Conclusion

This 9th lecture has led in essence to a negative result. No links can be observed between phytochrome (P_{fr}) and important cell functions which would enable the development of a molecular hypothesis of P_{fr} action in the control of cellular growth. However, the discovery by SCHOPFER and OELZE [265] of a threshold mechanism in control of hypocotyl lengthening by $P_{fr\ (ground\ state)}$ has opened a new and fascinating possibility of understanding in modern terms one of the oldest problems of plant physiology: control of growth.

Selected Further Reading

LOCKHART, J. A.: The analysis of interactions of physical and chemical factors on plant growth. Ann. Rev. Plant Physiol. *16*, 37 (1965).
MOHR, H.: Biologie als quantitative Wissenschaft. Beilage zu: Naturwiss. Rdsch., Heft 7, 779 (1970).
ROTH, K., LINK, W., MOHR, H.: RNA- and protein synthesis in connection with control of longitudinal cell growth. Cytobiologie *1*, 248 (1970).
STEINER, A.M.: Rasch ablaufende Änderungen im Gehalt an löslichen Zuckern und Zellwandkohlenhydraten bei der phytochrominduzierten Photomorphogenese des Senfkeimlings (*Sinapis alba* L.). Planta *82*, 223 (1968).

Modulation of Hypocotyl Longitudinal Growth by $P_{fr\ (ground\ state)}$ through a Threshold Mechanism

This lecture is mainly based on recent investigations by P. SCHOPFER and H. OELZE-KAROW [265]. Their work contributes to growth physiology as well as to a better understanding of the phytochrome system *per se*.

The longitudinal growth of the hypocotyl of the mustard seedling is inhibited by the action of phytochrome [159]. This response has been used to elaborate a quantitative relationship between the amount of physiologically active $P_{fr\ (ground\ state)}$ and the extent of the response, Δm.

$$\Delta m = f\ ([P_{fr\ (ground\ state)}])$$

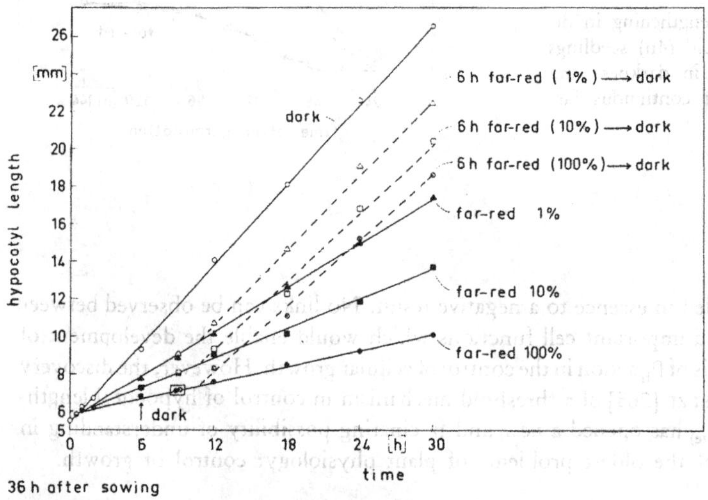

Fig. 123. The growth rate of the mustard hypocotyl depends on irradiance (100% = standard far-red source). Irrespective of irradiance, the time lag (Δt) between the termination of light and the termination of growth inhibition is always approximately 4.8 h (after [265])

In the course of the 4th lecture a distinction was made between the action of $P_{fr\ (ground\ state)}$ and P_{fr}^{*}, and the conclusion was drawn that under continuous irradiation the effect of light is predominantly (or even exclusively) due to some excited species of P_{fr} (P_{fr}^{*}), while the effect of P_{fr} in the dark is due to $P_{fr\ (ground\ state)}$.

1. Experimental Data

In light-mediated inhibition of hypocotyl lengthening in the mustard seedling both modes of action of P_{fr} can be demonstrated. Fig. 123 shows some growth kinetics of the hypocotyl of intact mustard seedlings. In the dark the growth rate is constant, at least between 36 and 66 h after sowing. Continuous standard far-red light (350 µW · cm^{-2} = 100 per cent) leads to an 80 per cent reduction in growth rate. Under lower irradiances (one-tenth or one-hundredth of the standard) linear growth kinetics are also observed; however, the growth rates are higher. There is a strictly logarithmic relationship between decreasing growth rate and increasing irradiance. This is a characteristic of the "high irradiance response" [99, 264]. In the 4th lecture the "irradiance dependency" of the "high irradiance response" was explained by the assumption that the relative concentration of P_{fr}* is a function of the total absorption by the phytochrome species. A fascinating new observation (Fig. 123) is that when the seedlings are irradiated for 6 h with far-red light of 100, 10 or 1 per cent of standard irradiance, and are then returned to darkness, growth will continue at the established far-red rate for about 4.8 h. By this time the growth rate abruptly increases to a rate similar to that of the dark controls. In what follows the symbol "Δt" will be used for the duration of time between the termination of the irradiation and the termination of the growth inhibition. In the present example (Fig. 123) the value of Δt is about 4.8 h. Since Δt is obviously independent of the previous irradiance, one is tempted to assume that the value of Δt is determined by the amount of $P_{fr\ (ground\ state)}$ which is left in the seedling after the light is turned off. We recall from the 1st and 4th lectures that this amount (P_{fr}/P_{tot}) depends on wavelength but does *not* depend on irradiance. The data in Fig. 123 support the assumption that Δt might be a function of the amount of $P_{fr\ (ground\ state)}$ which is left in the seedling after the termination of the "high irradiance" irradiation. Fig. 124 shows that Δt is independent of the duration of a far-red preirradiation. When the far-red light is turned off after 5 min, 6 h, 12 h, or 18 h the subsequent growth kinetics are always consistent with $\Delta t = 4.8$ h. Furthermore, the data indicate that the term Δt is independent of the developmental stage of the seedling within the period of investigation although the growth rate in the dark obviously decreases with increasing duration of the far-red preirradiation.

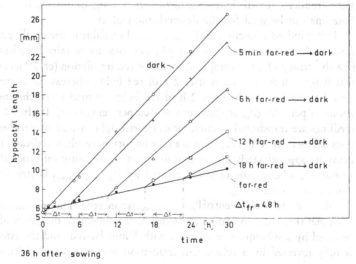

Fig. 124. The kinetics of mustard hypocotyl lengthening in complete darkness (–o–), continuous far-red light (–•–) and in darkness following a pretreatment with 5 min, 6 h, 12 h, or 18 h of far-red light (after [265])

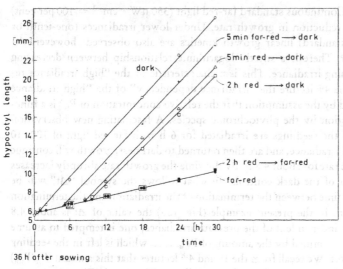

Fig. 125. Some irradiation programs to demonstrate the threshold response of hypocotyl lengthening. After 5 min far-red light (-+-) the Δt value is 4.8 h; after 5 min red light (-△-) Δt is 8.5 h; after 2 h red light (-□-) Δt is 7.2 h. If 2 h red light are followed by continuous far-red light (-■-), there is no difference as compared to the continuous far-red kinetics (-●-) (after [265])

If Δt is indeed related to the amount of $P_{fr \text{ (ground state)}}$ which is left in the seedling at the beginning of the dark period, Δt should depend on the wavelength of the preceding irradiation in a predictable manner. This is indeed the case (Fig. 125). While 5 min of far-red light lead to a Δt value of 4.8 h, 5 min of red light lead to a Δt value on the order of 8.5 h. In these experiments continuous irradiation with standard far-red and red light produced growth kinetics which were not significantly different. Thus, a common base-line can be used for the determination of Δt.

If Δt is indeed a function of $[P_{fr \text{ (ground state)}}]$ available in the seedling at the onset of darkness, a substantial reduction to total phytochrome by sustained red irradiation must result in a shortening of Δt as compared to a brief red irradiation (cf. 7th lecture). Fig. 125 shows that indeed Δt is only 7.2 h after 2 h of red light, whereas the corresponding Δt value after 5 min red is 8.5 h. After 2 h of red light the total phytochrome has decreased by about 84 per cent (P_{tot} at time zero = 100 per cent) [149]. If after 2 h of red light, the seedlings are transferred to continuous far-red light instead of darkness, the growth rate does not increase (Fig. 125). This is a very important result. It indicates that a phytochrome decay by approximately 85 per cent does not significantly influence the "high irradiance response", whereas the response elicited by $P_{fr \text{ (ground state)}}$ is strictly dependent on the amount of available P_{tot}.

As far as the involvement of $P_{fr \text{ (ground state)}}$ is concerned, the operational criteria have clearly been met (Fig. 126). As far as the term Δt is concerned, the effect of 5 min red is fully reversed by a subsequent irradiation with 5 min far-red, and the effect of 5 min far-red is fully reversed by a subsequent irradiation with 5 min red light.

2. Theoretical Treatment

a) *The problem.* The problem was to correlate quantitatively the relative amount of P_{fr} (ground state) as determined by *in vivo* spectroscopy to the extent of a physiological response, in this case Δt. The function to be determined can be written as

$$\Delta t = f \left([P_{fr \text{ (ground state)}}] \right)$$

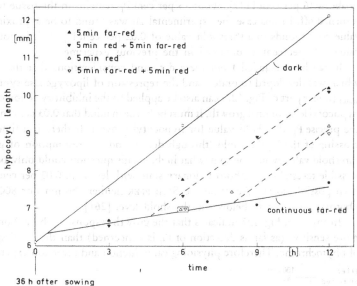

Fig. 126. Some experiments to demonstrate red, far-red reversibility. The irradiation with 5 min far-red (-▲-), 5 min red + 5 min far-red (-▼-), 5 min red (-△-) and 5 min far-red + 5 min red (-▽-) was performed at time zero (after [265])

b) *The assumptions.* Following BUTLER and associates [32] the photostationary state established by a brief saturating red irradiation is $P_{fr}/P_{tot} = 0.8$. This assumption is supported by a great number of experimental results, including *in vivo* measurements (cf. Fig. 10).

In the mustard seedling the decay of P_{fr} follows first order kinetics and is independent of the quantum flux density [149, 150]. The half-life of P_{fr} is close to 45 min at 36 h after sowing (25° C) [149].

A ratio of $P_{fr}/P_{tot} = 0.025$ is used as the photostationary state established by the standard far-red light source used. This value is derived from physiological data (P_{fr}-mediated repression of lipoxygenase synthesis in the mustard seedling [193]) and supported by HARTMANN's spectrophotometric measurements (cf. Fig. 10).

Finally the threshold concept is applied which was elaborated by OELZE-KAROW and associates [193] in connection with the phytochrome-mediated repression of lipoxygenase synthesis in the mustard cotyledons. According to this concept, which was described in the 7th lecture, the synthesis of lipoxygenase is controlled by a symmetrical threshold mechanism. Lipoxygenase synthesis is totally inhibited as soon as the relative concentration of P_{fr} is increased above the threshold value (1.25 per cent P_{fr}, based on $[P]_{tot}$ at time zero) and is resumed at full speed as soon as the relative concentration drops below this value (cf. Fig. 91).

c) *The conclusions* (Fig. 127). If 80 per cent P_{fr} is established by 5 min red light and if the P_{fr} pool decays with a half-life of 45 min, a P_{fr} concentration of about 0.03 per cent (based on P_{tot} at time zero) is reached after 8.5 h (the experimental Δt). If 2.5 per cent P_{fr} is established by 5 min of far-red light and if the P_{fr} pool decays with a half-life of 45 min, a P_{fr} concentration of 0.03 per cent will be reached after 4.8 h. This coincides with the experimental Δt. If the amount of P_{tot} is reduced by a 2 h red irradiation down to about 16 per cent [149], about 13 per cent P_{fr} remains in the tissue when the red light is turned off. In this case the experimental Δt was found to be approximately 7.2 h. This value corresponds to a threshold value of 0.02 per cent P_{fr}. This is not too far from the value 0.03 per cent as obtained in the previous experiments.

It can be concluded from these data and considerations that the threshold concept which was developed to understand the repression of lipoxygenase synthesis by $P_{fr\ (ground\ state)}$ (upper part of Fig. 127) can also be applied to the inhibitory action of $P_{fr\ (ground\ state)}$ on hypocotyl longitudinal growth. It must be borne in mind that 0.03 per cent is not necessarily the precise P_{fr} threshold value for hypocotyl growth. If there is a time lag between the crossing of the P_{fr} threshold (through decay) and the resumption of growth, the actual threshold value would be somewhat higher. This question could only be settled if it were possible to establish a photostationary state with less than 0.03 per cent P_{fr}. This has not been possible so far. According to SCHOPFER neither 756 nm nor 800 nm establishes a photostationary state below the threshold level [261].

In any case, Fig. 127 indicates that the growth response is about 2 orders of magnitude more sensitive (as far as detection of P_{fr} is concerned) than the usual spectrophotometry of phytochrome. Therefore physiological induction and reversion experiments can poten-

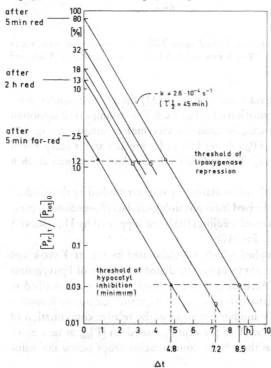

Fig. 127. A kinetic model to illustrate the threshold regulation by $P_{fr\ (ground\ state)}$ with respect to lipoxygenase synthesis and hypocotyl lengthening in the mustard seedling. The model is based on the following assumptions: (1) The decay of a given P_{fr} pool in the dark follows a first order kinetics whereby the half-life is 45 min. (2) Standard red light establishes a photostationary state with 80 per cent P_{fr}; standard far-red light etablishes a photostationary state with 2.5 per cent P_{fr}. The experimental values of Δt indicate a threshold value of 1.2 per cent P_{fr} in the case of the lipoxygenase repression and 0.03 per cent P_{fr} in the case of hypocotyl inhibition. This is a *minimum* value. For details see text (after [265])

tially be performed under conditions where no P_{fr} can be measured spectrophotometrically. This explains at least some of the so-called "phytochrome paradoxes" [109].

d) *A summary* (Fig. 123). The data can be best explained by the assumption that two independent phytochrome responses are involved. Firstly, the "high irradiance response" which controls the growth rate under continuous irradiation through $[P_{fr}^*]$. Secondly, a threshold response in which only $P_{fr\ (ground\ state)}$ is involved. This mechanism controls the growth rate in the dark period following irradiation.

3. Growth-limiting Proteins

Let us touch briefly on the question of how the threshold control by $P_{fr\ (ground\ state)}$ could perhaps be understood in terms of molecular events at the protein level. Studies with *Avena* coleoptile segments have indicated [38] that the as yet unidentified growth-limiting proteins (GLP) are functionally unstable. After protein synthesis has been inhibited the GLP pool decays within 20-25 min. The assumption (Fig. 128) that synthesis of a similarly unstable GLP is inhibited by P_{fr} (in the ground state) through a threshold mechanism would help to explain the facts observed in the mustard hypocotyl, including the short but consistent time lags before growth is terminated after the onset of light (cf. Fig. 65) or resumed after the threshold value has been passed in the downwards direction. Recent experiments indicate [261] that there might be a time lag in the latter case of on the order of half an hour.

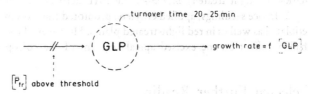

Fig. 128. This is a tentative model to explain the action of $P_{fr\ (ground\ state)}$ on hypocotyl lengthening in terms of growth-limiting proteins (GLP). It is assumed that $[P_{fr\ (ground\ state)}]$ blocks the synthesis of GLP as long as it is above the threshold concentration. As soon as $[P_{fr\ (ground\ state)}]$ drops below the threshold value, GLP synthesis can proceed at full speed. The regulation of protein synthesis by $P_{fr\ (ground\ state)}$ is almost instantaneous (e. g. Fig. 92)

4. A Final Comment

Threshold regulation of hypocotyl lengthening requires the perfect synchronization of all those cells which are involved in hypocotyl lengthening. This is true for the longitudinal as well as for the azimuthal dimension. The challenging question of how such an integration can be achieved cannot be answered so far. We shall return to this problem in the course of lecture 21.

Appendix: On the Use of Excised Segments in Phytochrome Research

This is a difficult matter, not only in phytochrome research. The real problem is that the inherent difficulties in using excised segments and in interpreting the data obtained in terms of the intact system, are by no means always recognized by the investigators. A further problem arises from the fact that most investigators have performed end-point determinations only. Information about the kinetics of the response is in general not available. Such data are difficult to evaluate. It must be emphasized again that in most cases kinetic data must be obtained. The results which were discussed in the course of this 10th lecture make the point very clearly. End-point determinations alone are not sufficient and may actually be misleading in the interpretation of physiological processes (e. g. [40]). Three examples:

1. In etiolated rice plants, the growth of intact coleoptiles was inhibited by an exposure to red light, while that of excised segments in a buffer medium was promoted by irradiation with red light, and both red light effects were reversed if the red irradiation was immediately followed by an irradiation with far-red light [201].

2. ENGELSMA and van BRUGGEN [59] have published data which indicate clearly that there is no correlation between ethylene production and phenylalanine ammonia-lyase (= PAL) development in gherkin hypocotyl segments. They further found that ethylene did not increase PAL activity in intact hypocotyl tissue but did so in excised tissue. Levels of ethylene which caused maximum inhibition of PAL synthesis in intact seedlings (both dark- and light-treated) *increased* the PAL activity in 2 mm hypocotyl segments.

3. In rice seedlings, applied ethylene promoted the growth of intact coleoptiles in totally etiolated as well as in red light-treated plants. However, the same concentration of ethylene had no effect on the excised apical segment of the coleoptile [114].

Selected Further Reading

CLELAND, R.: Instability of the growth-limiting proteins of the *Avena* coleoptile and their pool size in relation to auxin. Planta *98*, 1 (1971).

SCHOPFER, P., OELZE-KAROW, H.: Demonstration of a threshold regulation by phytochrome in the photomodulation of longitudinal growth of the hypocotyl of mustard seedlings (*Sinapis alba* L.). Planta *100*, 167 (1971).

The Problem of the Primary Reaction of Phytochrome

The term "primary reaction of phytochrome" is used to designate the first reaction in which P_{fr}, the physiologically active species of the phytochrome system, becomes involved after its formation. This reaction can be written as

$$P_{fr} + X \xrightarrow{\text{yields}} P_{fr} X$$

This formula is not very satisfactory, for at least three reasons: Firstly, there is no way so far to measure this reaction directly, that is, by physical methods. One must depend entirely on physiological information. Secondly, it is very probable that P_{fr} can act from the ground state (that is, from a non-excited state) as well as from some excited state, designated by P_{fr}^*. The so-called "high irradiance responses" (HIR) cannot be explained so far without the assumption of P_{fr}^* (cf. 4th lecture). Our discussion will be restricted to the primary reaction of P_{fr} in the ground state, that is, operationally speaking, to P_{fr} acting in the dark. Thirdly, this formula implies as a matter of course, that X (the primary reaction partner or reaction site of P_{fr}) is the same in all photoresponses mediated by P_{fr}. Thus this model does not account for the *specificity* of response which is so obvious in phytochrome-mediated control of plant growth and development. For some reason I do not fully understand, it is generally believed that the primary reaction of P_{fr} is the same in all cells (Fig. 129, lower part). With the same confidence, however, one can maintain that, even in the primary reaction, P_{fr} acts differently in different cells or in different compartments of a particular cell (Fig. 129, upper part). The truth can only be ascertained by experimentation. It is obvious that a decision must be made between these two alternatives before any progress with respect to the molecular details of the formula $P_{fr} + X \rightarrow P_{fr} X$ can be expected. Let me describe briefly some experimental results which invariably seem to indicate that the *upper* alternative must be favored. I shall proceed from weak to stronger arguments.

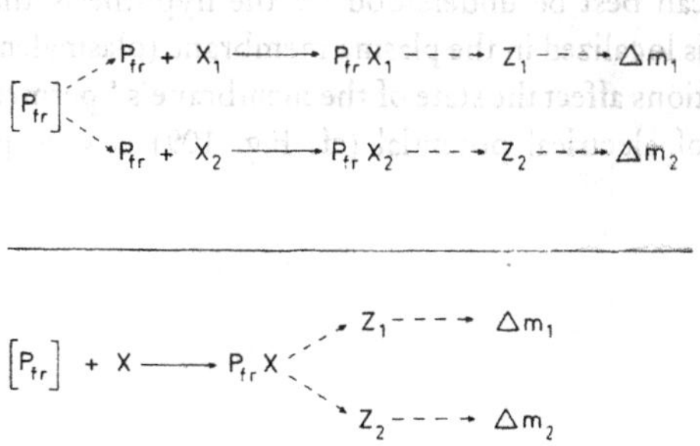

Fig. 129. Two alternative models of the primary reaction partner (or site) of P_{fr} (ground state). The lower model is not consistent with the facts presently available. While other, more complicated, models might be conceived to explain the available experimental information, the upper model has the advantage of being simple as well as consistent with the facts. [P_{fr}], P_{fr}-pool of the system; $X_{1, 2}$, reaction partner(s) or site(s) of P_{fr}. Nothing is implied about the degree or type of difference between the different Xs; $Z_{1, 2}$, intermediate(s) between $P_{fr} X$ and the responses; Δm, extent of P_{fr}-mediated responses (in general, change of rates)

$$[P_{fr}] \begin{cases} P_{fr} + X_1 \longrightarrow P_{fr} X_1 \dashrightarrow Z_1 \dashrightarrow \Delta m_1 \\ P_{fr} + X_2 \longrightarrow P_{fr} X_2 \dashrightarrow Z_2 \dashrightarrow \Delta m_2 \end{cases}$$

$$[P_{fr}] + X \longrightarrow P_{fr} X \begin{cases} Z_1 \dashrightarrow \Delta m_1 \\ Z_2 \dashrightarrow \Delta m_2 \end{cases}$$

1. Different P_{fr} Populations

SCHÄFER and associates [18a, 249] have claimed that in cotyledon and nook tissue of mustard or pumpkin seedlings more than one population of P_{fr} can be detected with respect to the reaction rate of phototransformation. If this concept can be defended against serious objections [241], it would be relevant for our present considerations, since there is no reason to assume that these photochemically different P_{fr} populations are located in the same compartment and are engaged in the same primary reaction. Rather, the different reaction rates would indicate a different interaction between the P_{fr} molecules and their molecular environment.

2. Polarotropism of Fern Sporelings [61]

This phenomenon has been dealt with in the 3rd lecture. Let us recall the situation with respect to phytochrome at the tip of the apical cell of a fern sporeling (cf. Fig. 35). The phytochrome molecules (whose axis of maximum absorption is indicated by dashes) are oriented in a specific dichroic manner. All results which have been obtained with this system can be interpreted only if one assumes (a) that the tip of the apical cell will grow at the point where most P_{fr} is present and (b) that the axis of maximum absorption of phytochrome turns by 90° during the transition of the red-absorbing form P_r to the far-red-absorbing form P_{fr}, and *vice versa*. The important point at the moment is that P_{fr} exerts a strictly local effect within the apical cell. P_{fr} determines precisely the location of the growing point of the cell. The effect of P_{fr} is not distributed or communicated all over the cell. Furthermore, the effect of P_{fr} does not seem to be related to the nucleus or to any other organelle within the cell [65].

On the other hand, in many phytochrome-mediated responses, e.g. phytochrome-mediated anthocyanin synthesis, a large part of the cell is very probably implicated, including control at the level of transcription and translation (cf. Fig. 159). It will be shown in the 13th lecture that phytochrome-mediated anthocyanin synthesis in the mustard seedling very probably involves a *simultaneous* action of P_{fr} at the level of transcription and translation (cf. Fig. 144). In this connection a report by GALSTON [74] is of considerable interest. He found, using a microspectrophotometer, that in sections of etiolated oat and pea tissue typical phytochrome signals (difference spectra) could only be obtained from the nuclear envelope. The size of the signal indicates that in these tissues the nucleus must be the site of localization of most of the cellular phytochrome. On the other hand though, the "Tanada effect" as observed in tips of secondary roots of mung beans (cf. 8th lecture) can best be understood by the hypothesis that a considerable phytochrome population is localized in the plasma membrane (plasmalemma) and that the phytochrome transformations affect the state of the membrane's "permeability". The immediate cause of the changes of electrical potential (cf. Fig. 109) is very probably ionic movement [187].

3. Threshold Regulation of Lipoxygenase Synthesis *vs.* Phytochrome-mediated Anthocyanin Synthesis as a Graded Response

While regulation of lipoxygenase synthesis by $P_{fr\ (ground\ state)}$ in the cotyledons of the mustard seedling requires a threshold concept for its explanation (cf. Fig. 91), P_{fr}-mediated anthocyanin synthesis in the *same* system does not [145]. The operational criteria for the involvement of $P_{fr\ (ground\ state)}$ in mediation of anthocyanin synthesis are fulfilled perfectly (cf. Fig. 11). However, there is no indication of a threshold for P_{fr} down to very low levels of P_{fr} [145]. Fig. 130 illustrates the principal difference between regulation of lipoxygenase synthesis and anthocyanin synthesis in the mustard seedling. While a cooperative model for the primary reaction of P_{fr} with respect to lipoxygenase synthesis must be considered (cf. 7th lecture), no threshold regulation is detectable in the case of anthocyanin synthesis and thus no cooperative model for the primary reaction of P_{fr} is required.

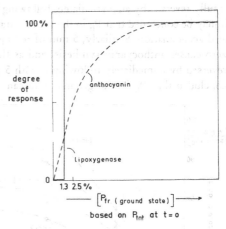

Fig. 130. This sketch illustrates the difference between regulation of lipoxygenase synthesis and anthocyanin synthesis in the same system (cotyledons of mustard seedlings). While the regulation of lipoxygenase follows a rigorous threshold pattern (——), the $[P_{fr}]$-response curve for anthocyanin synthesis very probably extrapolates to zero (based on data from OELZE-KAROW et al. [193] and LANGE et al. [145])

4. Control by $P_{fr\ (ground\ state)}$ of Extension Growth and Anthocyanin Synthesis in One and the Same Cell (cf. Fig. 68)

Lengthening of the hypocotyl of a mustard plant is predominantly due to cellular lengthening, at least between 36 and 72 h after sowing. Therefore organ lengthening represents lengthening of the average hypocotyl cell (cf. 10th lecture). Lengthening of the hypocotyl cells is controlled by P_{fr} through a threshold mechanism very similar to the one described for lipoxygenase (cf. 10th lecture). Therefore the respective primary reaction of P_{fr} (in the ground state) must have some high degree of cooperativity, that is, the primary reaction of P_{fr} with respect to cell lengthening requires the simultaneous cooperation of several or even many P_{fr} molecules at a control center (cf. 7th lecture). On the other hand, in anthocyanin synthesis in the same hypocotyl cells, no threshold control by P_{fr} could be detected. To repeat: lengthening of a subepidermal hypocotyl cell is controlled by P_{fr} through a threshold mechanism, whereas phytochrome-mediated anthocyanin synthesis in the same cell is not.

The arguments we have been discussing so far were derived from experiments which were not done primarily for the purpose of contributing to our understanding of the primary reaction of P_{fr}. Now I am going to describe some experiments which were intentionally designed to be relevant to this problem.

5. Control by P_{fr} of Ascorbic Acid and Anthocyanin Accumulation in the Mustard Seedling

Let us first compare some details of phytochrome-mediated ascorbic acid and anthocyanin accumulation in the mustard seedling with respect to the question of whether or not the primary reaction of P_{fr} (in the ground state) is the same in both responses.

5 min of red light (which establishes a phytochrome photoequilibrium of about 80 per cent P_{fr}) applied to a dark-grown seedling at time zero will transiently increase the rate of ascorbic acid accumulation (Fig. 131). Because the effect of 5 min of red light is fully reversed by an immediately following irradiation with 5 min of far-red light [254], it may be concluded that P_{fr} in the ground state can effectively control the rate of ascorbic acid accumulation. Similarly, 5 min of red light applied to a dark-grown seedling at time zero causes anthocyanin synthesis, and as the effect of 5 min of red light can be fully reversed by immediately following it with 5 min of far-red light (cf. Fig. 11), it may be concluded that P_{fr} in the ground state can cause anthocyanin synthesis.

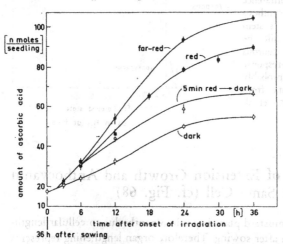

Fig. 131. Time-course of ascorbic acid accumulation under continuous far-red light, continuous red light and after one brief irradiation (5 min) with red light given at time zero (that is, 36 h after sowing) (after [171])

Against this background, the following experiment can be performed (Fig. 132): continuous far-red light which probably acts exclusively through P_{fr}^{*} exerts a strong effect on anthocyanin synthesis. Between 6 and 18 h after the onset of light a nearly constant rate of anthocyanin accumulation is maintained. However, even after prolonged treatments with continuous far-red light, the anthocyanin-producing system of the mustard seedling still responds to changes of the P_{fr} level (exerted by 5 min of red light after the termination of far-red light); an increase of $P_{fr\ (ground\ state)}$ from about 2.5% P_{fr} (photoequilibrium characteristic for far-red light) to about 80% P_{fr} (photoequilibrium characteristic for red light) leads to an increase in the rate of anthocyanin synthesis.

Corresponding experiments for ascorbic acid accumulation lead to a different result (Fig. 133). The ascorbic acid-accumulating system of the mustard seedling does not respond at all to changes of the level of $P_{fr\ (ground\ state)}$ (exerted by 5 min of red light after the

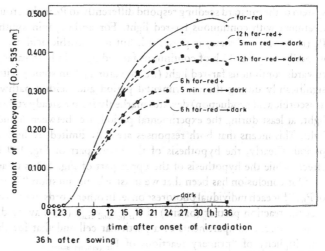

Fig. 132. Time-courses of anthocyanin accumulation under continuous far-red light and in experiments where the standard far-red light was turned off 6 or 12 h after the onset of irradiation. The samples were either placed in the dark immediately or received 5 min red light before the transfer to darkness (after [171])

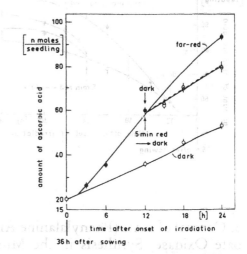

Fig. 133. Time-course of ascorbic acid accumulation under continuous far-red light and in an experiment where the standard far-red light was turned off 12 h after the onset of irradiation. The samples were either placed in the dark immediately or received red light for 5 min before the transfer to darkness (after [171])

termination of far-red light). A program of 12 h far-red light plus 5 min red light → dark leads to the same result as a program of 12 h far-red light → dark. Obviously the system can no longer respond to P_{fr} in the ground state, while the sensitivity towards continuous far-red (that is, towards P_{fr} in some excited state) is not significantly decreased.

The question arises as to whether or not the sensitivity of the system towards P_{fr} in the ground state can be restored by a dark period. This is indeed the case (Fig. 134). After a dark period of 12 h (following 12 h of continuous far-red light) the ascorbic acid-accumulating system of the mustard seedling again responds to $P_{fr \text{ (ground state)}}$ (5 min red → dark).

As a result the anthocyanin-synthesizing system and the ascorbic acid-accumulating system of the mustard seedling respond differently to P_{fr} in the ground state after a prolonged treatment with continuous far-red light. For anthocyanin synthesis, sensitivity towards P_{fr} (in the ground state) is preserved, but for ascorbic acid accumulation this sensitivity is totally lost after 12 h of continuous far-red light. In both cases, however, the sensitivity towards continuous far-red light (i.e. towards P_{fr}^*, in some excited state) does not change significantly during the experimental period and neither pathway (anthocyanin as well as ascorbic acid production) deviates markedly from a steady state under continuous far-red light, at least during the experimental period, i.e. between 6 and 18 h after the onset of light. This means that both responses are only limited by light during our experimental period. Clearly, the hypothesis of the lower part of Fig. 129 contradicts the observed effects while the hypothesis of the upper part of Fig. 129 is in accordance with the facts.

Our conclusion has been that we must ask the question "What is the primary reaction of P_{fr}?" for each individual photoresponse. In other words, we must adhere to the hypothesis that the reaction partner (or site) of $P_{fr \text{ (ground state)}}$ (X) may be different in different cells or in different compartments of a particular cell and that for this reason we can expect a multiplicity of "primary reactions of $P_{fr \text{ (ground state)}}$".

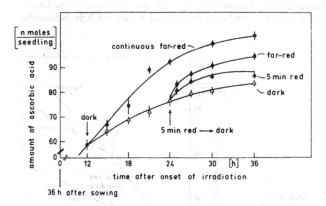

Fig. 134. Time-course of ascorbic acid accumulation under the program: 36 h dark (time zero) – 12 h far-red light – 12 h dark – 5 min red light ⟶ dark. The corresponding control kinetics are given also: 12 h far-red light ⟶ dark (dark); 12 h far-red light – 12 h dark ⟶ far-red light (far-red); continuous far-red light (after [171])

6. Control of PAL (Phenylalanine Ammonia-lyase) and AO (Ascorbate Oxidase) Synthesis in the Mustard Seedling

The results obtained with ascorbic acid and anthocyanin accumulation have been confirmed at the enzyme level. A detailed comparison of the two enzymes AO and PAL has led to the principal result (cf. Table 1,2) that both enzymes respond to short-term irradiation with red or far-red light if the brief treatment is given to a dark-grown seedling. However, a 5 min treatment with red, far-red or dark-red light after 12 h of continuous far-red light only leads to a positive result in the case of PAL; AO does not respond significantly within 6 h after the short-term treatments (Table 12). To summarize at this point of the discussion, the experimental information obtained so far leads to the conclusion that there is no model available for a single and universal primary reaction of P_{fr} and probably never will be. Obviously a multiplicity of models is needed for the primary reactions of P_{fr}.

Table 12. The action of $P_{fr \text{ (ground state)}}$ in the mustard seedling after a prolonged far-red treatment. 5 min red light establish a P_{fr}/P_{tot} ratio on the order of 0.8, 5 min black-red (756 nm) establish a P_{fr}/P_{tot} ratio < 0.01. Standard far-red light maintains a P_{fr}/P_{tot} ratio of about 0.03.
PAL = Phenylalanine ammonia-lyase
AO = ascorbate oxidase (after [49], [52])

Treatment after Sowing (d = dark; fr = far-red)	PAL Activity $\left[\dfrac{\text{p moles trans-cinnamic acid}}{\text{min} \cdot \text{pair of cotyledons}}\right]$
36 h d + 12 h fr	66 ± 2
36 h d + 12 h fr + 3 h fr	104 ± 2
36 h d + 12 h fr + 3 h d	66 ± 4
36 h d + 12 h fr + 5 min red + 3 h d	75 ± 3
36 h d + 12 h fr + 5 min 756 nm + 3 h d	58 ± 3
	AO Activity $\left[\dfrac{\text{nmoles ascorbate}}{\text{min} \cdot \text{pair of cotyledons}}\right]$
36 h d + 12 h fr	77 ± 3
36 h d + 12 h fr + 6 h fr	132 ± 4
36 h d + 12 h fr + 6 h d	112 ± 5
36 h d + 12 h fr + 5 min red + 6 h d	112 ± 5
36 h d + 12 h fr + 5 min 756 nm + 6 h d	108 ± 3

7. General Conclusions

It may be concluded from the previous section that the action of P_{fr}^* can change the sensitivity of a system towards $P_{fr \text{ (ground state)}}$. After termination of the continuous far-red light the sensitivity of the system towards $P_{fr \text{ (ground state)}}$ eventually recovers, but it always takes some time, even in the case of PAL, whereas for about half an hour after the onset of dark the system does not respond significantly to differing levels of $P_{fr \text{ (ground state)}}$ (Fig. 135). Only after a considerable period of time is the sensitivity towards $P_{fr \text{ (ground state)}}$ fully restored. These results have led to the further conclusion that during the action of continuous standard far-red light (i. e. P_{fr} in some excited state), the responding systems are in general insensitive towards $P_{fr \text{ (ground state)}}$.

8. Tentative Models (Fig. 136)

Let me discuss briefly two simple and formal models for the primary reaction of $P_{fr \text{ (ground state)}}$ which we have been using over the years as a guide in our efforts to approach this problem experimentally.

The first model (Fig. 136, upper part) uses P_{fr} as the effector molecule proper. It describes the situation if the primary binding sites of P_{fr} (X_i) were sites along a chromosome (transcrip-

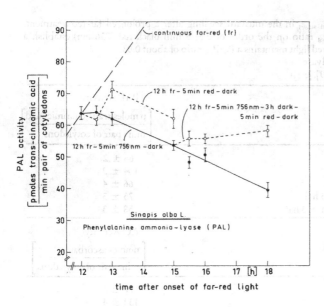

Fig. 135. The effect on PAL synthesis of a brief irradiation (saturating with respect to the photostationary state of the phytochrome system) after termination of a 12 h treatment with continuous standard far-red light. 5 min red establish a photostationary state with approximately 80 per cent P_{fr}; 5 min black-red (756 nm) establish a photostationary state with less than 1 per cent P_{fr} (unpublished data from K. PETER and H. MOHR)

Primary Reactions of P_{fr}

Model 1: 3 elements, P_r P_{fr} X_i (binding site)

postulates :

(1) $P_{fr} + X_i \rightleftharpoons P_{fr} X_i$ $k_{c(1)}$ very large

(2) $P_r + X_i \rightleftharpoons P_r X_i$ $k_{c(2)}$ very small

Model 2: 6 elements, P_r P_{fr} X_i membrane, E_e , E_i
                                                    (E = effector proper)

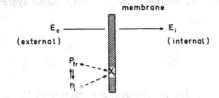

postulates :

(1) $P_{fr} X_i$ hight permeability for E

(2) $P_r X_i$ low permeability for E

(3) E_i must have a high turnover

Fig. 136. Two formal models for the primary reaction of P_{fr} (ground state). These deliberately simple models have been developed for the purpose of guiding experimental efforts until more sophisticated models become feasible (for details see text)

tional control) or at a ribosome (or polysome) (translational control). The model requires only 3 elements and 2 postulates about the binding forces (or equilibrium constants). The primary reaction of P_{fr} would be both rapid and fully reversible.

The second model (Fig. 136, lower part) involves a change of "membrane permeability" (including "active transport") by P_{fr}. The concept that P_{fr} may act on membrane permeability is favored by some workers in the field but unfortunately only in very general and sometimes even nebulous terms (cf. 8th lecture). Our model requires at least 6 elements and 3 postulates, and is obviously much more complex than the first model.

It would require only a slight modification of the second model to account for the hypothesis that cAMP (cyclic adenosine monophosphate) can act as a "second messenger" [217, 124] in some P_{fr}-mediated responses. One can imagine that cAMP would be produced at the membrane through the action of adenylate cyclase and rapidly destroyed inside the compartment under the influence of cAMP phosphodiesterase. Both processes, synthesis as well as destruction, can possibly be controlled by P_{fr}. Recent findings [121] indeed suggest that the level of cAMP in the mustard seedling is regulated by phytochrome.

Let us recall just one experimental result to indicate the order of magnitude of the time constants which are involved in "molecular" responses mediated by $P_{fr\ (ground\ state)}$. Fig. 137 reminds us that lipoxygenase synthesis is rapidly arrested as soon as the threshold level of $P_{fr\ (ground\ state)}$ is exceeded. Furthermore, lipoxygenase synthesis is resumed immediately and at full speed as soon as the threshold value is passed in the other direction. After 2 h of continuous red light, the level of total phytochrome is so low (about 16 per cent of P_{tot} at time zero) that a shift from red to far-red light will lead to a P_{fr} value far below the threshold value. Clearly, increase of lipoxygenase is resumed immediately after the shift from red to far-red light. We have preferred to explain these and similar data in terms of the first model (translational control) since we felt that the rapidity of the response in both directions can hardly be reconciled with the second model.

The phenomenon of "escape from control by P_{fr}" has also been used to determine the rapidity of irreversible action of P_{fr} once it is formed. If one separates a short irradiation with red light from the corresponding irradiation with far-red light by a dark period, it turns out the reverting effect of the far-red light decreases with increasing length of the dark period (e.g. [21]). While in lettuce seed germination (cv. Grand Rapids) the time

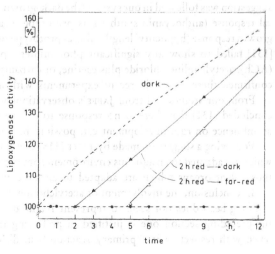

Fig. 137 Some experimental results obtained in connection with repression of lipoxygenase synthesis by $P_{fr(ground\ state)}$ (cf. 7th lecture) which demonstrate the *rapidity* of the threshold response in both directions (that is, inhibition as well as resumption of lipoxygenase synthesis). The relevant experiment (2 h red $\longrightarrow$ far-red) is described in the text (after [193])

period needed to decrease reversibility by 50 per cent is on the order of several hours, in the case of flower initiation in *Chenopodium rubrum* the far-red light is already without reverting effect after 70 min. In flower initiation in *Xanthium pennsylvanicum* this period of time is only 40 min and in *Chenopodium album* only 1 min. Finally, in *Pharbitis nil*, the escape of the response from control by P_{fr} occurs on the order of seconds (70, 105).

Should one want to uphold the doctrine of one single primary reactant of $P_{fr\ (ground\ state)}$ the following model is at hand: X is always and everywhere the same (e.g. it is a particular allosteric protein, altered by P_{fr}), but it can be located at different sites in a cell, e.g. in the plasma membrane or in the nucleus. This means that X is embedded in different molecular neighborhoods (sites), and due to the interaction with its neighbors will have somewhat different properties (e.g. different binding constants for P_{fr}). Even if X is the same everywhere, the system [X plus site] is not and thus the primary reaction is not the same even if X (as an isolated substrate) were the same everywhere. As an example, X could be a constituent of the plasma membrane, or X could be a repressor protein sitting on a gene locus.

9. The Significance of Acetylcholine

JAFFE [117] reported that acetylcholine given for 4 min in the dark was able in dark-grown mung bean seedlings to substitute for red light (i.e. for P_{fr} in the ground state) in reducing the formation of secondary roots, inducing increased H^+ efflux, and causing the root tip to adhere to a negatively-charged glass surface. Since eserine, which is a specific inhibitor of acetylcholine-esterase, inhibited the far-red-mediated release from the glass surface and since acetylcholine is present in mung bean roots and the levels of extractable acetylcholine are controlled by phytochrome in the usual manner (operational criteria), JAFFE concluded that acetylcholine may act as a *local hormone* which regulates these phytochrome-mediated phenomena. He further concluded that these findings possibly provide an explanation of the mechanism by which photoconversion of the phytochrome holochrome might be coupled to the morphogenic responses.

Since the "primary reaction" of P_{fr} is still a matter of controversy [240, 171] JAFFE's suggestion was followed in our system (the dark-grown mustard seedling) using a biochemical response (anthocyanin synthesis as prototype of a "positive" response) as well as a growth response (hypocotyl lengthening as prototype of a "negative" response). The results [131] failed to show any significant photomimetic properties of acetylcholine-chloride (ACh), acetylcholine-chloride plus eserine, or carbamylcholine-chloride (CCh). JAFFE has confirmed these results in recent experiments with mustard seedlings [118].

From our studies and from JAFFE's observations with the mustard seedling [118], we concluded [131] that there is no response to ACh or CCh in the shoot system, while an influence on root development can possibly be detected under certain circumstances.

Following a suggestion made by JAFFE [118] one could argue that plants or plant organs which are adapted to an aqueous environment sometimes respond to ACh (or CCh), while those plant organs which are adapted to an air environment do not.

In conclusion, the involvement of acetylcholine in the action of P_{fr} in secondary roots of mung bean does not seem to represent *the* mode of action of P_{fr}. As documented in the preceding section one is justified in preferring the concept that P_{fr} acts differently (even with respect to its "primary reaction") in different systems and even within the same cell. It is not excluded, of course, that the level of acetylcholine may be controlled in many plant tissues by phytochrome [93] and that acetylcholine may act as a "transmitter" or "local hormone" in *some* phytochrome-mediated responses like the ones studied by Jaffe. However, recent results obtained with the fungus *Trichoderma viride* indicate that the photomimetic effect of acetylcholine is *not specific* for phytochrome-mediated responses. GRESSEL et al. [85] report that the blue-light-mediated induction of conidiation can be mediated through an acetylcholine-type response, in a similar manner to that observed in JAFFE'S experiments.

10. NAD Kinase and Phytochrome

A new idea came on the scene when TEZUKA and YAMAMOTO [270] reported that NAD kinase activity is controlled by phytochrome. It was found with seedlings of light-grown *Pharbitis nil* (cv. Violet) that a brief irradiation with red light in the middle of the night period will raise the level of NADP. The operational criteria for the involvement of phytochrome are fulfilled. It was further found that the NAD kinase activity in an extract (supernatant) of *Pharbitis* cotyledons was increased after a red treatment. The most important fact reported was that there is a phytochrome control of NAD kinase in a partially-purified phytochrome preparation from pea stem tissue. The K_M of NAD kinase for NAD is decreased from 1.84×10^{-3} M in the dark to 0.90×10^{-3} M in red light. Two possibilities were considered for an explanation of these findings: (1) Phytochrome itself has an NAD kinase activity. (2) Phytochrome is closely associated with NAD kinase. It was thought that P_{fr} may contribute to the decrease of the K_M value of the NAD kinase reaction by causing a conformational alteration of NAD kinase. Unfortunately no further reports on this subject have appeared so far.

Appendix: On the Mechanism of the "High Irradiance Response" (that is, the Mechanism of P_{fr}^* Action)

While P_{fr} in the dark acts from the ground state (S_0) ($P_{fr_{S_0}} + X \rightarrow P_{fr} X$), it has been postulated [94, 264] that under continuous light (that is, under "high irradiance conditions") P_{fr} will act from some excited state (P_{fr}^*). Unfortunately the nature of this excited state is not known. The most important characteristic of the "high irradiance response" is the dependence on irradiance (cf. 4th lecture).

We may assume that P_{fr} in the ground state is a singlet as are essentially all pigments of importance in biological systems. Therefore our writing of the primary reaction as

$$P_{fr_{S_0}} + X \rightarrow P_{fr} X$$

is justified. As far as the "high irradiance response" is concerned we must recall the principle that photochemical reactions may occur from any excited electronic state. However, a further distinction should be made [202] between those reactions which occur in thermally equilibrated vibrational levels of an electronically excited state (equilibrium reactions) and those reactions which occur from vibrationally excited levels, either of the ground state or of an electronically excited state, before vibrational relaxation has occurred. HARTMANN

[94] has advanced an explanation of the dependence on irradiance of the "high irradiance response" as occurring from a vibrationally excited level. However, those reactions which occur from non-equilibrium vibrational levels must have rate constants greater than 10^{11} to 10^{13} s^{-1} and will, therefore, be unimolecular only, except for molecules at very close separations in highly-concentrated systems [202]. Since bimolecular reactions (like P_{fr}^* + X → P_{fr} X) almost universally occur from the lowest electronically-excited state (ES$_1$ or ET$_1$ type), it seems probable that the species of phytochrome acting in the "high irradiance response" is either $P_{fr_{S_1}}$ or $P_{fr_{T_1}}$. Since the steady state concentration of the active species is a function of irradiance, the irradiance dependency of the "high irradiance response" is understandable. Unfortunately, nothing is known about the excited states of P_{fr}. While fluorescence of the P_r form has been measured with an excitation maximum near 670 nm and an emission maximum near 690 nm, no fluorescence could be detected for P_{fr} [106]. It must further be assumed that P_{fr}^* is not used up in the reaction. After being de-excited ($P_{fr_{S_0}}$) every molecule can be used repeatedly. Therefore it is better to write the reaction as

$$h \cdot \nu \quad \begin{matrix} P_{fr}^* & X \\ \{ \ \} & \sim & \{ \\ P_{fr_{S_0}} & X' \end{matrix} \to \text{response}$$

On the other hand, $P_{fr_{S_0}}$ can act only once (or a multiple but *fixed* number of times) with its primary reactant. This is clearly indicated by the strict validity of the reciprocity law for the $P_r \to P_{fr}$ reaction. A given number of quanta yields a given effect irrespective of irradiance if the longest time of irradiation does not exceed the order of 5 min [145]. Even for physiological reasons it seems advisable to keep the two reactions

$$P_{fr_{S_0}} + X \to P_{fr} X$$

and

$$h \cdot \nu \quad \begin{matrix} P_{fr}^* & X \\ \{ \ \} & \sim & \{ \\ P_{fr} & X' \end{matrix}$$

totally apart [265]. Phytochrome is involved in both reactions but everything else seems to be different. This conclusion is almost inevitable if one tries to understand physiological responses like control of lipoxygenase synthesis [193], where it has only been possible to detect the action of $P_{fr\ (ground\ state)}$ so far, or control of hypocotyl lengthening [265] where the investigators were forced by the facts to conclude "that the two photoreactive systems" ... (that is, $P_{fr\,(ground\ state)}$ and P_{fr}^*)... "involve separate populations of phytochrome which inhibit cell lengthening by independent control mechanisms".

Suggested Further Reading

GALSTON, A.W., SATTER, R.L.: A study of the mechanism of phytochrome action. In: Recent Advances in Phytochemistry (1971).
JAFFE, M.J.: Evidence for the regulation of phytochrome-mediated processes in bean roots by the neurohumor, Acetylcholine. Plant Physiol. 46, 768 (1970).

MOHR, H., BIENGER, I., LANGE, H.: Primary reaction of phytochrome. Nature (Lond.) 230, 56 (1971).

SCHOPFER, P., OELZE-KAROW, H.: Demonstration of a threshold regulation by phytochrome in the photomodulation of longitudinal growth of the hypocotyl of mustard seedlings (Sinapis alba L.). Planta 100, 167 (1971).

SMITH, H.: Phytochrome and photomorphogenesis in plants. Nature (Lond.) 227, 665 (1970).

Interaction between Phytochrome and Hormones

Fig. 138 illustrates two simple alternatives with respect to the question of whether or not an interaction between P_{fr} and hormones exists.

Even in recent research papers and review articles the hypothesis has repeatedly been advanced as a matter of course that the mechanism of P_{fr} action involves plant hormones at some intermediate step (left side). To quote GALSTON and DAVIES in a recent review article in Science [75]: "When the plant is illuminated with red light, transforming the red absorbing form of phytochrome to the far-red absorbing form, kaempferol synthesis is increased in stems, and the resultant increase in indoleacetic acid oxidase activity could account for the decline in stem growth rate after red irradiation. In leaves, the same red light induces the formation of a related 3', 4'-dihydroxy flavonoid, quercetin, which serves as an inhibitor of indoleacetic acid oxidase and could account for the promotion of leaf growth". The fact is that any rigorous attempts to identify plant hormones as intermediates between P_{fr} and the photoresponses have failed so far, in negative as well as in positive photoresponses. We must visualize the alternative, namely, that P_{fr} and the plant hormones do not interact. Indeed, the available information indicates that at least in the range of regulation of enzyme synthesis, P_{fr} and hormones act as independent factors (right side of Fig. 138).

Let us first discuss an example where the non-existence of interaction between a hormone (GA_3) and P_{fr} is obvious.

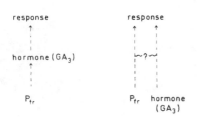

Fig. 138. Left, in this alternative it is assumed that the action of P_{fr} involves plant hormones at some intermediate step; right, in this alternative it is assumed that P_{fr} and plant hormones do not interact (at least in the range of regulation of enzyme synthesis). Our first example (amylase induction) favors the concept on the right, whereas our second example (peroxidase induction) possibly supports the model on the left

1. Induction of Amylase by P_{fr}

It is well known that the hormone GA_3 (gibberellic acid) will induce *de novo* synthesis of amylase in aleurone layers of embryo-free half-caryopses of some cereal grains like barley or wheat [67, 115]. On the other hand, in the cotyledons of the mustard seedling amylase can be induced in the usual way by the active form of phytochrome, P_{fr}. Table

Table 13. Conventional induction-reversion experiments to demonstrate the involvement of phytochrome (P_{fr} in the ground state) in the control of the increase of amylase in the mustard seedling. The light treatment was given 36, 42, 48 and 54 h after sowing (25° C). Amylase was assayed 60 h after sowing (after [53])

Treatment after Sowing	Amylase Activity	
	$\left[\dfrac{-\ \mu g\ starch}{min\ \cdot\ seedling}\right]$	%
36 h dark	4.6	100
36 h d + 24 h d	8.1	176
36 h d + 4 × (5 min red + 355 min d)	18.8	409
36 h d + 4 × (5 min red + 5 min fr + 355 min d)	13.8	300
36 h d + 4 × (5 min fr + 355 min d)	13.6	296

dark = d; far-red = fr

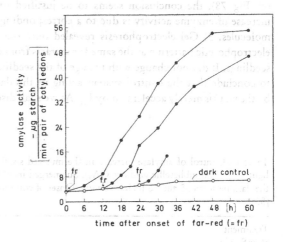

Fig. 139. Time-course of amylase increase in the cotyledons of the mustard seedling in darkness and under the control of continuous far-red light. Onset of light: 36, 48 and 60 h after sowing. For the kinetical studies, only the cotyledons were used since they contain most of the enzyme and the degree of induction is higher than in the rest-seedling (hypocotyl plus taproot) (after [53])

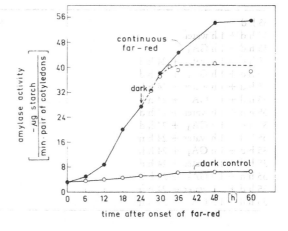

Fig. 140. Time-course of amylase activity in mustard cotyledons after termination of far-red light (dark). Let us recall the background of the experiment: The physiological effectiveness of a given photostationary amount of P_{fr} depends on the irradiance of the far-red light (cf. 4th lecture). If one turns off the far-red light, the physiological effectiveness of the remaining $P_{fr\ (ground\ state)}$ very probably drops to nearly zero instantaneously (cf. 11th lecture) (after [53])

13 shows that the conventional operational criteria for the involvement of P_{fr} (ground state) in amylase induction are verified, namely, complete reversal of a brief induction with red light by immediately following with a brief irradiation with far-red light. In Fig. 139 some induction kinetics are shown for continuous far-red light, i.e. for P_{fr}^* under steady state conditions. The far-red $\rightarrow$ dark kinetics, one of which is shown in Fig. 140, indicate that in the case of amylase too, enzyme decay does not play a significant role during the experimental period. Furthermore, it is obvious that the presence of P_{fr} is continuously required to maintain a high rate of amylase synthesis.

Some characteristics of the mustard amylase can be summarized as follows [53]: 1) Amylase can be induced by P_{fr} in the cotyledons as well as in the rest-seedling. However, most of the enzyme is localized in the cotyledons. 2) Application of inhibitors of RNA and protein synthesis (cf. Fig. 70) at low doses leads to a strong inhibition of amylase increase. The results obtained with amylase coincide *quantitatively* with those for PAL in mustard seedlings. Since in the case of PAL *de novo* synthesis of the enzyme has been demonstrated in the mustard cotyledons by means of density labelling with deuterium (cf. Fig. 78), the conclusion seems to be justified that even in the case of amylase the increase of enzyme activity is due to a corresponding increase in the number of enzyme molecules. 3) Gel electrophoresis revealed only one intense and two weak bands. The electrophoretic pattern was the same for extracts from dark-grown and from far-red-grown seedlings. It did not change with the age of the seedling. Howeyer, it would be premature to conclude that the control system acting in the dark (which is completely unknown, by the way) is merely accelerated by P_{fr}. As we shall discuss later (16th lecture) in connection

Table 14. Control of amylase increase in the mustard seedling by phytochrome (continuous far-red light = fr) and gibberellic acid (= GA$_3$). Submerged incubation in GA$_3$ or water was performed in the dark between 35 and 36 h after sowing. Onset of continuous far-red light: 36 h after sowing (after [53])

Treatment after Sowing (dark = d)	Concentration of GA$_3$ [$\mu g \cdot ml^{-1}$]	Amylase Activity $\left[\dfrac{-\mu g\ starch}{min \cdot seedling} \right]$
35 h d	-	4.5
35 h d + 1 h water	-	4.6
35 h d + 1 h GA$_3$	50	4.6
35 h d + 1 h water + 24 h d	-	10.8
35 h d + 1 h GA$_3$ + 24 h d	50	10.0
35 h d + 1 h water + 24 h fr	-	51.5
35 h d + 1 h GA$_3$ + 24 h fr	50	36.2
35 h d + 1 h water + 24 h d	-	10.5
35 h d + 1 h GA$_3$ + 24 h d	100	7.8
35 h d + 1 h water + 24 h fr	-	48.8
35 h d + 1 h GA$_3$ + 24 h fr	100	52.8
35 h d + 1 h water + 24 h d	-	10.8
35 h d + 1 h GA$_3$ + 24 h d	0.35	10.9
35 h d + 1 h water + 24 h fr	-	50.0
35 h d + 1 h GA$_3$ + 24 h fr	0.35	35.8

with PAL, the control system acting in the dark and the control exerted by P_{fr} seem to be totally independent although the PALs formed in darkness and in the light show the same characteristics in gel electrophoresis and other tests (cf. Fig. 79). 4) The efforts to classify the mustard amylases as α- or β-amylases (following the conventional criteria) did not lead to unambiguous results. In any case, when mustard amylases were incubated with starch the pattern of products of hydrolysis was similar to that produced by commercially-available barley β-amylase and was not similar to that produced by *Bacillus subtilis* α-amylase. The term amylase is used in order to signify the total amylolytic activity of the mustard cotyledons.

The main problem has been whether or not GA_3 can induce amylase in the mustard seedling and whether or not P_{fr} can induce amylase in the embryo-free half-caryopses of barley and wheat. Table 14 gives the principal results: When mustard seedlings were incubated with solutions of GA_3 at concentrations which are saturating for cell elongation in the mustard hypocotyl (cf. Fig. 111), there was no induction of amylase. Rather, P_{fr}-mediated induction of amylase was significantly lower in GA_3-treated seedlings in comparison to the water controls, and this negative effect was already detectable at low concentrations. While this non-specific effect can probably be understood in terms of competition between hypocotyl and cotyledons for common pools of metabolites, the positive statement can be advanced that GA_3 does not participate in the mediation of amylase induction by P_{fr} in the cotyledons of the mustard seedling.

To summarize the principal results of our investigations: Synthesis of amylase in the embryo-free half-caryopsis of wheat and barley is induced by GA_3. An influence of continuous far-red light cannot be detected. Synthesis of amylase in the cotyledons of the mustard seedling can be induced by P_{fr}. Any positive influence of GA_3 cannot be detected. Obviously the amylase enzyme which consists of several isoenzymes can be induced in different systems by different effector molecules. There is no detectable interaction between P_{fr} and GA_3 as far as induction of amylase is concerned.

2. Induction of Peroxidase by P_{fr}

A photoresponse of the mustard seedling has recently been investigated[1] which shows a completely different kinetic pattern: the P_{fr}-mediated increase of total peroxidase activity[2]. The appearance of this enzyme activity in the axis of the seedling (hypocotyl plus taproot = rest-seedling) is not influenced by the presence of P_{fr} up to about 5-6 days after germination. (After this time the energy resources of the seedling are depleted and reliable results can no longer be obtained.) In the cotyledons, however, continuous far-red light induces a strong increase in peroxidase activity (Fig. 141). The conventional red-far-red induction reversion experiments indicate that the operational criteria for the involvement of P_{fr} in this response are fulfilled.

The ability of the cotyledons to produce peroxidase under the influence of P_{fr} is strongly dependent on the stage of development (Fig. 142): P_{fr} is effective only when it is formed

---

1 This part of the 12th lecture is based on recent data obtained by P. SCHOPFER and C. PLACHY. Permission to use this information in the present context is greatly appreciated.
2 The fact that several isoenzymes of peroxidase exist in the mustard seedling can be ignored in our present consideration.

before about 96 hours after sowing. However, peroxidase activity increases only after about 96 hours after sowing. In other words, the formation of P_{fr} in the cotyledons leads to enzyme synthesis in a period during which the competence to respond to P_{fr} has already been lost. Obviously the induction process is clearly separated in time from the realization

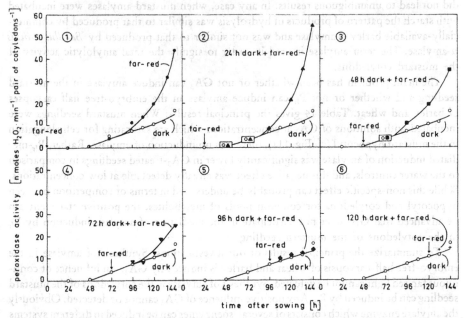

Fig. 141. Time-course of peroxidase activity in the cotyledons of mustard seedlings in the dark and under continusous standard far-red light. Onset of light: 0, 24, 48, 72, 96 or 120 h after sowing (after [262])

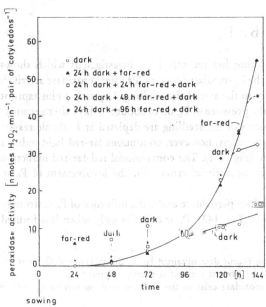

Fig. 142. Time-course of peroxidase activity in the cotyledons of mustard seedlings with different periods of far-red light. Onset of light: 24 h after sowing. Transfer to darkness: 48, 72 or 120 h after sowing (after [262])

of the response. Fig. 142 shows that formation of P_{fr} before 48 h after sowing is also ineffective in the induction of peroxidase activity. However, if the seedlings are irradiated in the proper period of time of "primary" differentiation (e.g. 24 until 72 h after sowing) and then transferred to darkness, peroxidase accumulates at the same rate as under continuous far-red light for at least $2^{1}/_{2}$ days. Similar results are obtained when the irradiation lasts for 48 hours longer before onset of darkness (120 h after sowing). The deviation of these far-red → dark kinetics from the far-red kinetics after about 130 h is probably an "artefact" due to re-etiolation of the seedlings in the dark which leads to a faster depletion of reserve materials from the cotyledons.

It appears from these data that the P_{fr}-mediated induction of an increased peroxidase synthesis is a virtually irreversible process, at least during the time in which the mustard seedling can be used for experimentation under our conditions. P. SCHOPFER has used the term "determination" (specifically "photodetermination") in contrast to the rapidly reversible "modulation" to characterize this type of response. However, it must again be emphasized that reversible modulation of a rate as well as determination of a rate will lead to an irreversible change in the amount of the product of a response only if the final product is stable (Fig. 143). Thus both "photomodulation" of the rate of enzyme synthesis and "photodetermination" will lead to "differentiation" only if the final product is stable.

As to the molecular mechanism, it is obvious that we need a more complicated model for the "photodetermination" than we need for the "photomodulation" of the rate of enzyme synthesis. Since P_{fr} rapidly disappears from the system in the dark, we need at

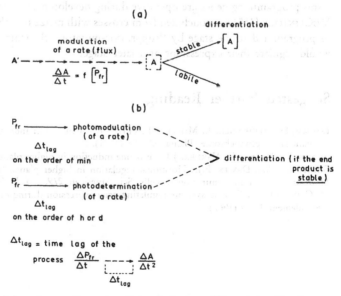

Fig. 143. (a) Modulation of a rate by an effector (e.g. P_{fr}) only leads to differentiation (that is, to an irreversible change in the system) if the end product of the metabolic chain (A) is stable (cf. Fig. 97). Rapid photomodulation of anabolic sequences is a widespread phenomenon in the mustard seedling, e. g. synthesis of anthocyanin, synthesis of ascorbic acid, " induction" of amylase or PAL synthesis, "repression" of lipoxygenase synthesis, control of hypocotyl lengthening. (b) Photomodulation of a rate by P_{fr}, as well as photodetermination of a rate by P_{fr}, can lead to differentiation if the end product of the anabolic sequence is stable

least one stable intermediate in the metabolic chain between P_{fr} and peroxidase synthesis which acts as a "transmitter" of the primary effect of P_{fr} to the peroxidase-synthesizing machinery. Apparently the "transmitter" can be formed in the presence of P_{fr} only at an early stage of "primary" differentiation (about 48-96 h after sowing) and can act only at a later stage of "primary" differentiation (about 96-120 h after sowing), resulting in an increased synthesis of peroxidase. A relatively stable mRNA or a hormone-like substance are possible candidates for this "transmitter". At any event it can be concluded from these considerations that P_{fr} can influence the differentiation of the mustard seedling on the level of enzyme synthesis in two different ways, by "photomodulation" and by "photodetermination", and that these two types of response cannot be produced by the same molecular mechanism.

To summarize the new aspects:

1. The phenomenon of photodetermination of *enzyme* synthesis might be a useful molecular model for the phenomena of determination in plant and animal embryology.

2. It is obvious that two "competences" must be kept apart:

   a) the competence of the seedling for P_{fr} with respect to transmitter synthesis (about 48 to 96 h after sowing).

   b) the competence of the mustard cotyledons for transmitter with respect to peroxidase synthesis (after 96 h).

3. A molecular model which would describe the effect of P_{fr} on peroxidase synthesis is not yet available. However, formal models which have been developed in classical genetics are obviously relevant. Genetic experiments with higher plants leave no doubt that mechanisms programming genes are operative during development of multicellular systems. B. McClintock [155] has concluded from crosses with maize that the action of genes could be programmed at one stage by "suppressor-mutator" elements (Spm) in a manner that would regulate their expressions at a later stage.

Suggested Further Reading

Drumm, H., Elchinger, I., Möller, J., Peter, K., Mohr, H.: Induction of amylase in mustard seedlings by phytochrome. Planta 99, 265 (1971).

Filner, P., Wray, J.L., Varner, J.E.: Enzyme induction in higher plants. Science 165, 358 (1969).

Galston, A.W., Davies, P.J.: Hormonal regulation in higher plants. Science 163, 1288 (1969).

Hadorn, E.: Transdetermination in cells. Sci. American 219, No. 5, 110 (1968).

McClintock, B.: Genetic systems regulating gene expression during development. Develop. Biol. Supplement 1, 84 (1967).

The Double Function of Phytochrome in Mediating Anthocyanin and Enzyme Synthesis

1. Anthocyanin Synthesis

In the mustard seedling anthocyanin synthesis can be mediated by phytochrome without the interference of any other photochemical mechanism [145]. However, the action of P_{fr} in the ground state (cf. Fig. 11) and of P_{fr}^* (cf. Fig. 52) must be kept separate. A significant finding has been that the duration of the "initial lag-phase" is constant (3 h at 25° C) for seedlings more than 30 h old and is specific for the system, being independent of the dose irradiance or quality of light (cf. Fig. 11, 52). The term "initial lag-phase" is used to designate the time-lag between the onset of light (applied to a seedling grown so far in complete darkness) and the first appearance of anthocyanin. The period of time over which actual production of anthocyanin occurs, can be called "production phase".

The finding that the length of the initial lag is a constant, specific for the system and independent of the dose, irradiance or quality of light is at least analogous to the reported fact [24], that the induction lag of the *E. coli* lac operon does not lengthen as the induction level is lowered. In other words, no change in induction lag with inducer level is found. Neither the time of first appearance of β-galactosidase in excess of the basal level nor the value of the intercept of the straight line part of the induction curve with the basal level is increased.

With respect to anthocyanin synthesis in the mustard seedling the following question arises: Is P_{fr} inactive in mediating anthocyanin synthesis during the 3 h of the initial lag?; in operational terms: Is an induction of anthocyanin synthesis by red light fully reversible for a period of 3 h? The answer is given by Fig. 144. Induction was performed with a standard red source. The reversion was done with 5 min of the wavelength 756 nm at an irradiance of 700 μW · cm^{-2} in order to keep the level of far-red induction as low as possible.

Complete (or nearly complete) reversal is only possible on the order of a few minutes after the onset of light. On the other hand, the rate of escape from reversibility (cf. 11th lecture) is neither rapid nor is the escape a total one, even after 4 h. This fact indicates that there is a continuous requirement for P_{fr} during the whole period of anthocyanin accumulation. In addition, P_{fr} clearly mediates anthocyanin synthesis during the lag-phase in spite of the fact that the actual synthesis of anthocyanin can proceed only after the lag-phase is overcome.

In the case of a secondary irradiation (e. g. 12 h light – 18 h dark: Light of the same quality and irradiance) no significant lag-phase can be detected (cf. Fig. 82). The same is true even if the secondary irradiation differs in quality and irradiance (cf. Fig. 105). The only requirement is that the rate of anthocyanin synthesis mediated by secondary irradiation does not exceed the one mediated by the primary irradiation.

These and further data [145] have led to the conclusion that during the initial lag-phase P_{fr} exerts two functions:

1) A potential for biosynthesis of anthocyanin ("capacity") is built up under the influence of P_{fr}. Since anthocyanin synthesis is very sensitive towards Actinomycin D during the lag-phase only [170] the assumption seems to be justified that long-lived mRNA required for anthocyanin synthesis is formed during the initial lag-phase. Thus, a tentative identification of the "capacity" with a given amount of stable mRNA has been made.

2) Anthocyanin synthesis is mediated by P_{fr} ($P_{fr \ (ground \ state)}$ as well as P_{fr}^{*}) but this mediation can only become effective after the "potential for biosynthesis" (i.e. long-lived mRNA) is eventually built up. Since anthocyanin synthesis is always very sensitive towards inhibitors of protein synthesis (cf. Fig. 73), it is thought that the second effect of P_{fr} on anthocyanin synthesis might be exerted on the level of translation (i.e. RNA-dependent

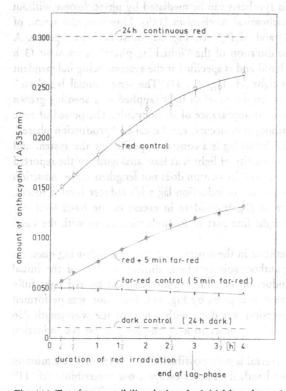

Fig. 144. Test for reversibility during the initial lag-phase. After irradiation of the mustard seedlings with standard red light of varying duration (abscissa) the seedlings were irradiated for 5 min with long wavelength far-red (756 nm; 700 µW ·cm^{-2}) and placed in the dark. Extraction of anthocyanin was performed 24 h after onset of the red light. The "red" controls received only red light of varying durations, the "far-red" controls only 5 min far-red each at the times indicated (after [145])

The data clearly show that the long-wavelength far-red (= black-red) which has been generated by a band interference filter (peak transmission at 756 nm, bandwidth 21 nm, irradiance 700 µW · cm^{-2}) rapidly and fully reverses a red-mediated induction. Further, we notice that there is a considerable induction effect exerted by this black-red itself. This fact means that the P_r absorbs considerable numbers of quanta transmitted by this interference filter. Since in extremely sensitive seed (*Nigella damascena*) germination could still be induced at 790 nm [147], it seems that P_r has a significant absorption even at longer wavelengths. However, since the light transmitted by the 756 nm filter acts like darkness in the lipoxygenase test ([190], cf. 7th lecture) the photostationary state established by this type of light must contain less than 1. 2 per cent P_{fr}

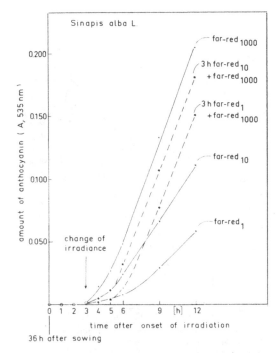

Fig. 145. The time-courses of anthocyanin accumulation in the mustard
seedling under continuous far-red light
in a typical "step-up experiment".
Three hours after the onset of far-red
the irradiances were increased by
100- or 1000-fold (dashed curves). "Far-
red$_{1000}$" denotes the standard far-red
irradiance (350 μW · cm^{-2}) (after
[145])

protein synthesis). P_{fr} is thought to increase the rate at which the protein-synthesizing
machinery runs.

These conclusions have been supported by a considerable number of further data, for
example the following: The time required to build up the "capacity" is determined by
the system. It is about 3 h under our standard conditions (25°C). The degree or extent
of capacity, however, is a function of the primary light treatment. For a "step-up experiment" (Fig. 145), that is, a change from a low to a high irradiance, there is a considerable
delay before the higher rate of synthesis becomes apparent. The additional time-lag seems
to be somewhat less than the initial lag, possibly for the reason that some "capacity"
was already built up under the influence of the lower irradiance. If the capacity for a
given irradiance is once built up it does not readily decay. We recall that there is no lag-phase
for anthocyanin (or enzyme) formation if the irradiance of the secondary exposure is not
changed to higher values (cf. Fig. 104). But even if the capacity is built up and maintained,
the rate of anthocyanin synthesis depends on irradiance (i.e. primarily on the steady state
amount of P_{fr}*). For "step-down experiments" (Fig. 146), the rate of anthocyanin accumulation adjusts within about 3 h to values similar to those mediated by the continuous lower
irradiances. This dependency on irradiance under conditions of continuous far-red irradiation is a characteristic of the "high irradiance response" (HIR). The HIR in the far-red
range can be fully explained in terms of phytochrome (cf. 4th lecture).

The conclusion that P_{fr} is continuously required during the production phase applies
to $P_{fr\ (ground\ state)}$ as well as to P_{fr}*. The fact that this continuous requirement can indeed
be satisfied (in part) by $P_{fr\ (ground\ state)}$ is shown by the data in Fig. 147. The rationale
is briefly as follows [145]: estimates can be made of the amount of $P_{fr\ (ground\ state)}$ present
in the mustard seedling at various times after a 5 min red irradiation given at time zero
and followed by either dark or 5 min far-red irradiation at some later time (half-life of

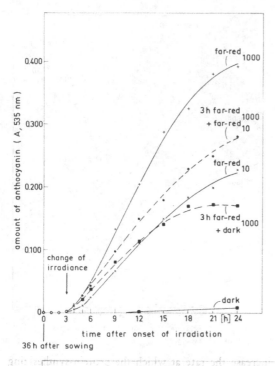

Fig. 146. The time-courses of anthocyanin accumulation in the mustard seedling under continuous far-red light in a typical "step-down experiment". Three hours after the onset of far-red the full irradiance was decreased by a factor of 100 or the light was turned off (dashed curves). "Far-red$_{1000}$" denotes the standard far-red irradiance (350 μW $\cdot$ cm^{-2}) (after [145])

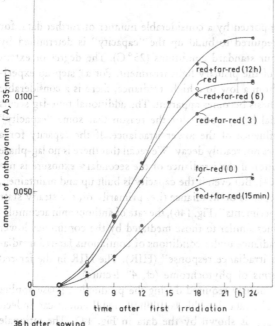

Fig. 147. The time-courses of anthocyanin accumulation in the mustard seedling after one or two brief irradiations (5 min each) with red or far-red light. Every sample (except the far-red control, fr (0), which received 5 min far-red instead) received 5 min red at time zero. The numbers in parentheses indicate the time when a second irradiation (5 min far-red) was given (after [145])

P_{fr} is 45 min at 25°C and 36 h after sowing). On the basis of the calculated data it could be predicted, for instance, that 5 min of far-red light, applied 12 h after the initial 5 min irradiation with red light, must have an inductive influence rather than a reverting influence on the time-course of anthocyanin synthesis. There is satisfactory agreement between prediction and experimental result (cf. Fig. 147). The important feature is the shift from reversion by far-red to *induction* by far-red light. This latter phenomenon becomes apparent 12 h after the red treatment.

2. Enzyme Synthesis

Phytochrome-mediated synthesis of phenylalanine ammonia-lyase (PAL) in the mustard seedling can be regarded as a model system for phytochrome-mediated enzyme synthesis (cf. 6th lecture). Every argument which has been used in the previous section with respect to anthocyanin synthesis applies to PAL synthesis as well. Let us recall the most important facts:

a) Fig. 81, as previously discussed, shows that one 5 min irradiation with red or far-red light, applied 48 h after sowing (time zero in the figure) will induce PAL synthesis in the mustard cotyledons. The effect is due to $P_{fr\ (ground\ state)}$. Continuous standard far-red light mediates a strong increase of PAL synthesis as well. This is a "high irradiance response". It is very probably due to P_{fr}^{*}. The important point for our present considerations is that the lag-phase after the onset of continuous light seems to be the same as in the case of a brief light treatment.

b) There is no detectable secondary lag-phase in PAL synthesis if the secondary irradiation is performed with the same light source as was used for the initial irradiation (cf. Fig. 82).

c) While the sensitivity of PAL synthesis towards inhibitors of protein synthesis is the same with primary and secondary irradiation [216, 49], this is not true for Actinomycin D. The sensitivity of the response towards this inhibitor is much less if applied before the secondary irradiation [216, 49]

d) Using the MAK-method of RNA separation and double-labelling of RNA with ^{14}C-uridine and ^{3}H-uridine, no phytochrome-mediated change at the level of RNA could be detected, even under conditions where P_{fr} rapidly mediates PAL-induction (12 h fr − 6 h d→ 20 min fr (or dark)) (cf. Fig. 96). While our failure to detect significant phytochrome-mediated changes at the RNA level in short-term experiments (secondary irradiation) could be due to an inadequate resolving power of the MAK separation technique, the more probable alternative is that the data of Fig. 96 mean that there is no change at the level of RNA during rapid enzyme induction by a *secondary* irradiation.

e) In the case of phytochrome-mediated repression of enzyme synthesis (lipoxygenase; cf. Fig. 137) there is no detectable time-lag. This means − if we interpret this fact in terms of gene regulation − that the life-time of the mRNA involved must be extremely short. In this case too, the more probable alternative seems to be that P_{fr} exerts its control on the level of translation while the synthesis of the pertinent mRNA is independent of phytochrome. However, in the case of lipoxygenase continuous RNA synthesis is required to allow enzyme synthesis [193].

f) The fact that in the case of PAL (cf. Fig. 102) as well as in the case of ascorbate oxidase (cf. Fig. 85), the primary (or initial) lag-phase will eventually disappear as the seedlings age, can possibly be explained by the assumption that the mRNA in question can eventually reach a non-limiting level even in the dark. This type of explanation is suggested by the fact that the elimination of the primary lag-phase is only completed after a stationary dark level of enzymatic activity has been attained (cf. Fig. 85, 102). Of course, enzyme synthesis in complete darkness requires the presence of the messenger involved. If the messenger has a considerable turnover some sensitivity towards Act.D must be expected (cf. Table 6).

g) A regulation of protein synthesis on the level of translation has recently been suggested for several eucaryotic systems, e.g. by HÄMMERLING [90] for *Acetabularia* and by TOMKINS et al. [272] for rat hepatoma cells. Further, it is well-known that during the early embryo development in many animals the only RNA which is translated is that which has already been accumulated in the egg cell (for references cf. [176a]).

In conclusion, we feel that the phenomena of phytochrome-mediated control of anthocyanin or protein synthesis can only be understood if we assume that P_{fr} can control transcription as well as translation. However, the formulation of molecular models to explain the mechanism of phytochrome action on the level of transcription and/or translation is still in its infancy (cf. 11th lecture). Those "controlling elements" which were detected by B. McCLINTOCK [155] in genetic experiments on anthocyanin synthesis in maize should probably be included in future models. It is possible that P_{fr} controls transcription through the action of the "suppressor-mutator element" (Smp).

Appendix 1: Function of P_{fr} in Ascorbic Acid Synthesis

Quite obviously, the hypotheses advanced for anthocyanin or enzyme synthesis do not apply without reservation to phenomena such as phytochrome-mediated modulation of ascorbate synthesis (cf. Fig. 98). This reponse does not depend on undisturbed RNA synthesis (cf. Table 9). Although the response requires protein synthesis which is to some extent undisturbed [12] there is no reason to assume that P_{fr} exerts its action on the level of translation. Furthermore, in the case of phytochrome-mediated ascorbate synthesis, the duration of the primary (or initial) lag-phase will *increase* with increasing age of the mustard seedling (cf. Fig. 101), a phenomenon not encountered so far in phytochrome-mediated enzyme synthesis. Finally, it has been concluded ([171]; cf. 11th lecture) that the "primary reaction" of P_{fr} cannot be the same in phytochrome-mediated anthocyanin synthesis and in phytochrome-mediated ascorbate synthesis. The two responses were compared under conditions where the initial lag-phase had been overcome. We suggest with respect to ascorbate synthesis that P_{fr} acts on the level of "membrane permeability", including an active transport of (a) rate-limiting metabolite(s).

Appendix 2: Induction of Nitrate-reductase in Corn Leaves

I shall refer only to those induction experiments in which etiolated plants (or plant parts) were used. In this case phytochrome seems to mediate the formation of the messenger RNA while the so-called "induction" of nitrate reductase by its substrate involves control of protein synthesis by nitrate. The situation (in our interpretation) is briefly the following [274]: Dark-grown oat and barley seedlings accumulate large amounts of nitrate in darkness but nitrate reductase activity does not increase above the low endogenous level until light is supplied (unfortunately only white light has been used so far in the experiments). However, when dark-grown oat leaves were "induced" in the light for 12 h and then returned to darkness, the activity continued to increase for another 24 h. The ability of corn leaves to produce an active nitrate reductase apparently depends on the presence of polyribosomes. In leaves from 10-day-old dark-grown seedlings the polyribosome level is low, and nitrate reductase cannot be induced in complete darkness, above the low endogenous level, regardless of nitrate availability. Polyribosomes are formed during the initial stages of a light treatment, and after 2 to 4 h after the onset of light nitrate reductase activity can be induced. The specific effect of light on the formation of polyribosomes may be related to the control of messenger RNA formation and concomitant monoribosome to polyribosome transformation. Recent results [274a] indicate that the ability of dark-grown shoots or leaves to form nitrate reductase in darkness decreases with increasing age. The loss of ability with age and the increasing lag period preceding enzyme formation in light are due to a loss of active polyribosomes as the seedlings age.

In conclusion, the data suggest that in the case of nitrate reductase the process of transcription is controlled by light (probably phytochrome) whereas the process of translation is controlled by nitrate. On the other hand, in the case of anthocyanin synthesis both processes (transcription and translation) are under the control of phytochrome.

Suggested Further Reading

LANGE, H., SHROPSHIRE, W., MOHR, H.: An analysis of phytochrome-mediated anthocyanin synthesis. Plant Physiol. 47, 649 (1971).
MOHR, H.: Regulation der Enzymsynthese bei der höheren Pflanze. Naturwiss. Rdsch. 23, 187 (1970).
MOHR, H., SITTE, P.: Molekulare Grundlagen der Entwicklung. München: BLV 1971.
TRAVIS, R.L., HUFFAKER, R.C., KEY, J.L.: Light-induced development of polyribosomes and the induction of nitrate reductase in corn leaves. Plant Physiol. 46, 800 (1970).

Repression of Lipoxygenase Synthesis by P_{fr}: The Problem of Primary and Secondary Differentiation

In this lecture I want to deal from an experimental point of view with a major problem of development which has been repeatedly emphasized by developmental biologists (e. g. [281, 247]) but is commonly neglected by molecular biologists familiar only with microorganisms: the problem of a double-action control mechanism in development of multicellular systems. A clear comprehension of the various aspects of this problem is necessary if speculations about the molecular mechanisms that control the epigenesis of higher forms [26] are to be consistent with the available facts.

In the 5th lecture we summarized the action of P_{fr} under the term "secondary differentiation (P_{fr})" and we called the process of differentiation which determines the cell's responsivity to P_{fr} "primary differentiation (P_{fr})". Returning to Fig. 61, we recall that the *specificity* of the cellular response must depend on the specific state of responsivity (or competence) of the cell and tissues at the moment when P_{fr} is formed in the seedling. It is obvious that we need a double-action control mechanism in order to understand even the simple situation sketched in Fig. 61.

Any satisfactory theory of development of multicellular systems must account not only for the static situation illustrated in Fig. 61, but also for the changes in time. The problem is whether or not the temporal sequences of *primary* and *secondary* differentiation can be separated, not only logically but also experimentally. It is postulated that the processes of P_{fr}-mediated *secondary differentiation* are fully programmed in the pattern of *primary differentiation*. It is further postulated that the process of *primary differentiation* is independent of phytochrome (Fig. 148). To check these postulates experimentally, one can make use of the fact that synthesis of lipoxygenase is controlled by P_{fr} in the mustard seedling.

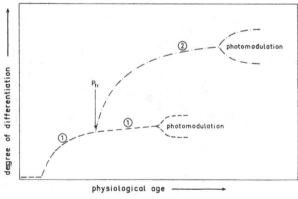

① : primary differentiation (P_{fr})

② : secondary differentiation (P_{fr})

Fig. 148. An illustration for the concept of the double-action control mechanism in the course of differentiation. Primary differentiation (with respect to P_{fr}) determines the cell's (or the system's) response to P_{fr}. The term secondary differentiation (with respect to P_{fr}) designates every developmental step which is triggered by P_{fr}. At the level of primary as well as at the level of secondary differentiation, photomodulations are possible

1. Threshold Regulation of Lipoxygenase Synthesis: A Recapitulation (cf. 7th Lecture)

Synthesis of the enzyme lipoxygenase in the cotyledons of the mustard seedling is controlled by $P_{fr\ (ground\ state)}$ through a threshold mechanism (cf. Fig. 91). The repression of enzyme synthesis by P_{fr} is a very rapid process after the threshold level is passed. Likewise, enzyme synthesis starts instantaneously and at full speed as soon as the P_{fr} level decreases below the threshold level (cf. Fig. 137). Thus P_{fr} rapidly inhibits synthesis of an enzyme and functions through an all-or-none control mechanism of high precision.

Since the threshold value is close to 1.25 per cent P_{fr} (based on P_{total} at time zero, that is, 36 h after sowing) and since standard far-red light establishes a photostationary state of the phytochrome system with approximately 2.5 per cent P_{fr}, standard far-red light immediately inhibits lipoxygenase synthesis, and the repression can be maintained by continuous far-red light over many hours (cf. Fig. 89). The question has been whether lipoxygenase synthesis responds to P_{fr} throughout the whole period of the seedling's development, that is, up to 84 h after sowing at 25°C.

2. Some Experiments to Validate the Concept of the Double-action Control Mechanism in Development [192]

Fig. 149 (open circles) shows the lipoxygenase kinetics in the dark. If one irradiates with standard far-red light from the time of sowing there is no control of lipoxygenase synthesis up to 33.25 h. At this point the full repression of lipoxygenase synthesis by far-red light (i.e., by a P_{fr} value above the threshold) suddenly comes into play, while at 48 h after sowing,

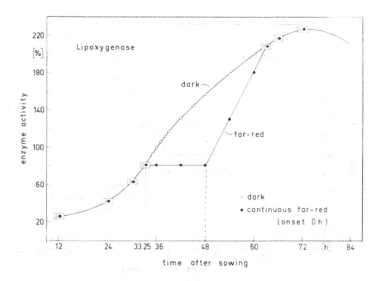

Fig. 149. The increase of lipoxygenase in the mustard seedling in continuous darkness and under continuous standard far-red light. The onset of far-red was at the time of sowing of the seeds (0 h). Where necessary, the following abbreviations are used: dark = d, far-red light = fr (after [192])

the seedling suddenly and completely escapes from control by P_{fr}. Enzyme synthesis is resumed even under continuous far-red light. Exactly the same temporal pattern is observed if the onset of continuous far-red light is at 24, 33.25 or 36 h after sowing (Fig. 150). Even in the latter case – onset of light at 36 h – the time of escape is at 48 h after sowing and the rate of enzyme synthesis after resumption is the same as with the other programs of irradiation. The kinetics after resumption return to the dark kinetics. Indications of an overshoot above the dark kinetics have never been found.

Many experiments have shown clearly that the system which produces lipoxygenase in the mustard cotyledons always escapes from the control by P_{fr} at 48 h after sowing (at 25° C and under standard conditions). The following series of figures contain the relevant experimental information. In Fig. 151 the increase of lipoxygenase in the mustard seedling is shown in continuous dark and under the influence of continuous standard far-red light. The onset of far-red light was 33.25, 36 or 42 h after sowing of the seeds. We note that under all circumstances the system escapes from control by far-red light at 48 h.

Spectrophotometric measurements of phytochrome performed with mustard cotyledons by E. SCHÄFER [250] indicate that neither at 33.25 h nor at 48 h are there any abrupt changes in the amount of detectable phytochrome (cf. Fig. 27). Furthermore, it can be concluded on the basis of the available data that up until 33.25 h as well as after 48 h, the level of P_{fr} in the far-red is well above the threshold level. Therefore it is improbable that the temporal pattern of response (no control up to 33.25 h, full control for 14.75 h and no control beyond 48 h) has anything to do with P_{fr}. This conclusion is supported by every experimental result obtained so far. Some further relevant data will now be discussed briefly.

Fig. 152 shows that the kinetics of lipoxygenase synthesis are identical under continuous

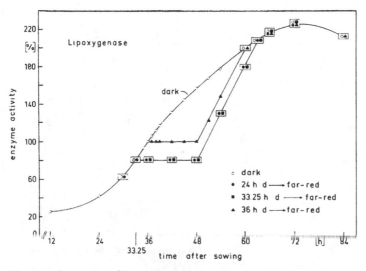

Fig. 150. The increase of lipoxygenase in the mustard seedling in continuous darkness and under the influence of standard far-red light. The onset of far-red was at 24, 33. 25 or 36 h after sowing of the seeds (after [192])

red and under continuous far-red light. Since the photostationary concentrations of P_{fr} as maintained by continuous red and far-red light differ considerably, this result clearly indicates that the only requirement with respect to P_{fr} is that the threshold level is passed. The actual amount of P_{fr} present in the cells is not important as long as this amount exceeds the threshold level.

The experiments for Fig. 152 were done nearly two years after the original experiments

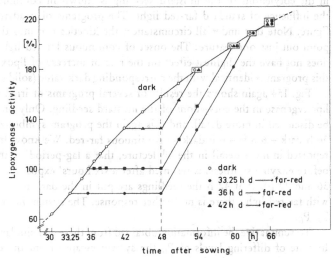

Fig. 151. The increase of lipoxygenase in the mustard seedling in continuous darkness and under the influence of standard far-red light. The programs of irradiation are indicated in the figure (after [192])

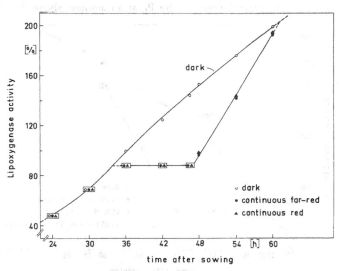

Fig. 152. The increase of lipoxygenase in the mustard seedling in continuous darkness and under the influence of continuous standard red or far-red light. Onset of light: at the time of sowing (after [190])

(in July-August 1971). It is obvious that the pattern of response has remained the same in principle. However, the onset of repression is somewhat later and the time of escape is slightly earlier. Thus, the duration of the time of sensitivity with respect to lipoxygenase repression has been shortened somewhat in the course of development of the seed material (kept at 4° C) over a period of about two years.

With Fig. 153 we return to the original data and again the increase of lipoxygenase in the cotyledons of the mustard seedling is shown in continuous darkness and under the influence of standard far-red light. The programs of irradiation are indicated in the figure. Note that under all circumstances the kinetics run into the dark kinetics. Let me point out just one feature: The onset of continuous far-red light at 48 h (solid squares) does not have the slightest effect on the rate of increase of lipoxygenase. Every value of this program is identical with the corresponding dark value (solid squares and open circles).

Fig. 154 again shows the results of several programs of irradiation on the increase of lipoxygenase in the cotyledons of the mustard seedling. Only one of these programs will be discussed in more detail, and it is again the program symbolized by the solid squares: 36 h dark – 6 h fr – 6 h dark → continuous far-red. We know from earlier experiments, reported in more detail in the 7th lecture, that a lag-period of about 45 min is observed before enzyme synthesis is resumed after a six hours' exposure to far-red light (between 36 and 42 h). If at 42 h the seedlings are put in the dark for 6 h and then reirradiated with far-red light, there is no further response. The system has escaped from the control by P_{fr}.

To summarize the information obtained from these and similar experimental programs: In spite of differing levels of P_{fr} the system escapes from the control by P_{fr} under all circumstances at 48 h after sowing.

It is concluded that the temporal pattern of response shown in Fig. 155 by the dashed line, must be determined by changes in the system on which P_{fr} acts and not by P_{fr}. Fig. 155 summarizes the experimental information regarding the control of lipoxygenase synthesis by far-red light, i.e. by P_{fr} at an amount above the threshold level. We note

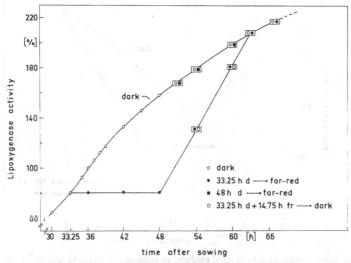

Fig. 153. The increase of lipoxygenase in the mustard seedling in continuous darkness and under the influence of standard far-red light. The programs of irradiation are indicated in the figure (after [192])

that the responsivity of the system towards P_{fr} changes abruptly in the manner of an all-or-none response. This is true for the onset (at 33.25 h) as well as for the termination (at 48 h) of the period of responsivity.

Two conclusions can be drawn with respect to Fig. 148: firstly, the repressive action of P_{fr} on lipoxygenase synthesis is obviously a function of primary differentiation. Secondly, the time-course of primary differentiation is independent of P_{fr}. The latter conclusion is

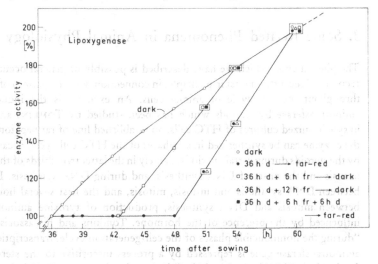

Fig. 154. The increase of lipoxygenase in the mustard seedling in continuous darkness and under the influence of standard far-red light. The programs of irradiation are indicated in the figure (after [192])

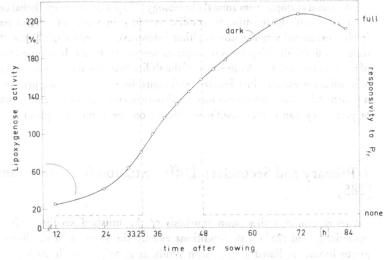

Fig. 155. A summary of the experimental information regarding the control of lipoxygenase increase by far-red light (i. e., by an amount of P_{fr} above the threshold level). Note that the responsivity of the system towards P_{fr} changes abruptly in the form of an all-or-none response (dashed line). This is true for the onset (33.25 h) as well as for the termination (48 h) of the period of responsivity (after [192])

based on several facts: a) Neither the beginning nor the end of the period of control (from 33.25 h to 48 h after sowing) are influenced by P_{fr} to any detectable extent. b) After resumption of synthesis the lipoxygenase kinetics are always parallel and run into the dark kinetics rather abruptly. In other words, as soon as the lipoxygenase kinetics reach the kinetics in continuous darkness, the control system acting in the dark takes over completely without any detectable after-effect of the light treatment.

3. Some Related Phenomena in Animal Physiology

The phenomenon which we have described is possibly of general occurrence. It has been recognized recently, at least in principle, in connection with the pattern of enzyme synthesis throughout the cell cycle of animal cells. An example is the induction of tyrosine aminotransferase by steroids which has been studied by TOMKINS and associates [272] in synchronized cultures of HTC cells, an established line of rat hepatoma cells. Although the enzyme can be synthesized in all phases of the HTC cell cycle, it can only be induced by the steroids during certain periods, namely in the latter two-thirds of the interval between mitosis and the onset of DNA synthesis, and during DNA synthesis. During the period between DNA synthesis and mitosis, mitosis, and the first several hours of the interval between mitosis and DNA synthesis, production of tyrosine aminotransferase is not influenced by the presence of the hormone. Tomkins and his associates conclude that "during the noninducible phases of the cell-generation cycle, transcription of the tyrosine aminotransferase gene is repressed by a process insensitive to the steroid, and that this repression is lifted during the inducible periods of the cycle". This is obviously another description for the hormone-independent regulatory process we have been calling "primary differentiation".

A second example from animal physiology [273]: Mammary epithelial cells of a pregnant mouse are initially sensitive to insulin, *in vitro*, in terms of several parameters, while freshly-explanted virgin tissue is either insensitive or only slightly sensitive. After some time in culture the virgin tissue acquires sensitivity to the hormone comparable to that of the pregnancy tissue. Acquisition of the ability to respond to insulin is largely independent of insulin. The events which lead to insulin sensitivity *in vivo* during pregnancy are probably not identical to those which produce sensitivity in virgin tissue *in vitro*. Thus, tissue isolation *or* pregnancy can confer insulin sensitivity on the mammary gland.

4. Primary and Secondary Differentiation in Anthocyanin Synthesis [285]

Let us return to anthocyanin synthesis of the mustard seedling. Under the standard conditions used (25° C, all conditions exactly the same as in the lipoxygenase studies) phytochrome-mediated anthocyanin synthesis is only possible 21 h after sowing and it ceases about 60 h after sowing independently of the irradiation program and independently of the amount of anthocyanin which has been accumulated (Fig. 156, 157). These facts clearly support the conclusion drawn previously with respect to lipoxygenase, that neither the beginning nor the end of the period of control by P_{fr} are to any detectable extent influenced by P_{fr}. This means that the time-course of primary differentiation is independent of P_{fr} (cf. Fig. 148).

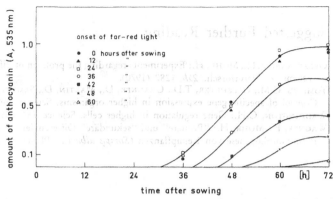

Fig. 156. The kinetics of anthocyanin accumulation in the mustard seedling under continuous standard far-red light. The time of onset of light is the experimental variable (after [285])

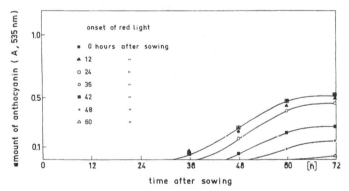

Fig. 157. The kinetics of anthocyanin accumulation in the mustard seedling under continuous standard red light. The time of onset of light is the experimental variable (after [285])

5. Further Approaches to the Problem

We cannot yet explain *primary differentiation* in terms of molecular biology. The causalities of primary differentiation are unknown in plants as well as in animals. But at least we can by now describe the problem clearly.

There are models which might be useful in further theoretical approaches to the problem, e.g. the model elaborated by TOMKINS and associates [272] or the model proposed by BRITTEN and DAVIDSON [26]. It is hoped that a clear recognition of the problem of primary and secondary differentiation on the level of enzyme synthesis will open the way towards an adequate *theoretical biology of development* and eventually lead to a molecular model of development in multicellular systems which is consistent with the facts of epigenesis.

A promising way to study the causalities of primary differentiation (in a model system) has recently been opened by the work of E. WELLMANN with cell suspension cultures of parsley [294]. This work will be described to some extent in the following lecture on light-mediated flavonoid biosynthesis as a biochemical model system for differentiation.

Suggested Further Reading

OELZE-KAROW, H., MOHR, H.: Experiments regarding the problem of differentiation in multicellular systems. Z. Naturforsch. *25b*, 1282 (1970).
TOMKINS, G.M., GELEHRTER, T.D., GRANNER, D., MARTIN, D., SAMUELS, H.H., THOMPSON, E.B.: Control of specific gene expression in higher organisms. Science *166*, 1474 (1969).
WADDINGTON, C.H.: Gene regulation in higher cells. Science *166*, 639 (1969).
WAGNER, E., MOHR, H.: "Primäre" und "sekundäre" Differenzierung im Zusammenhang mit der Photomorphogenese von Keimpflanzen (*Sinapis alba* L.). Planta *71*, 204 (1966).

Light-mediated Flavonoid Synthesis: A Biochemical Model System of Differentiation

1. The Starting Point [8]

The major processes taking place in an ordered, sequential manner during development of a multicellular system are growth, (cyto)-differentiation and morphogenesis. Growth involves the irreversible increase of certain parameters of the system (like length, weight, volume, DNA contents). Cytodifferentiation involves cellular specialization and occurs by the selective loss and acquisition of specific structures and functions. Morphogenesis involves the arrangement of cell populations in a precise and organized fashion in space and time.

Changes in state of differentiation in higher cell types are often mediated by simple external signals (or effectors), as, for example, in the action of hormones or embryonic inductive agents. P_{fr}, the active phytochrome, can be considered as being a prototype of such an effector. One could call the P_{fr} a "local hormone". We recall that in the case of phytochrome the effector molecules do not move from cell to cell (cf. Fig. 49). The P_{fr} molecules originate within a given cell from the inactive form (P_r) under the influence of light. The experimental advantages of phytochrome as compared to hormones are mainly the following: a) The serious problems which are caused by an external application of effector molecules in higher organisms do not arise. b) One can control the intracellular formation of P_{fr} precisely. c) One can measure the relative amount of P_{fr} actually present in the living cells by photometric methods of high precision. On the other hand, the principal information which can be obtained by the investigation of phytochrome-mediated responses very probably applies to hormones and embryonic inductive agents as well. From this point of view, phytochrome research is a field of *general* biology, including medicine.

In this 15th lecture the molecular basis of P_{fr}-mediated anthocyanin synthesis in the competent cells of a mustard seedling will be discussed as a model of a differentiation step in secondary differentiation. We shall further discuss a recent successful approach to the problem of primary differentiation using a cell suspension culture of parsley.

2. Phytochrome-mediated Anthocyanin Synthesis in the Mustard Seedling as a Model System for Secondary Differentiation

We recall the kinetics of anthocyanin accumulation in the mustard seedling under continuous far-red light (cf. Fig. 51). Formally we can describe the situation in terms of a transient modulation of the rate of anthocyanin synthesis. Since the end product is stable (at least within the time-period of experimentation), differentiation occurs (cf. Fig. 143). The problem was to describe in *molecular terms* the modulation of the rate of cyanidin synthesis. Although the mustard seedling forms five distinct anthocyanins, the aglycone is always cyanidin (Fig. 158).

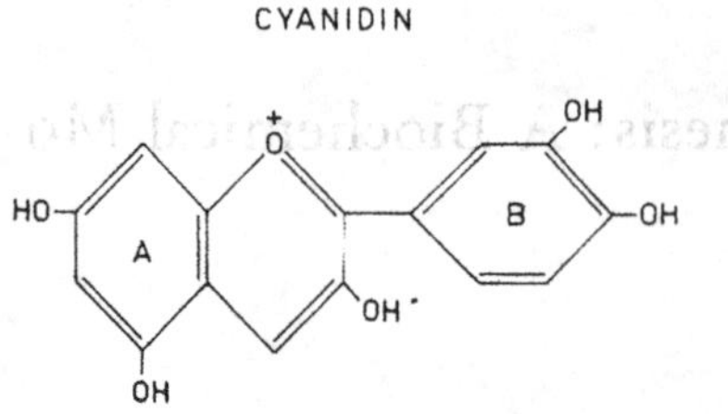

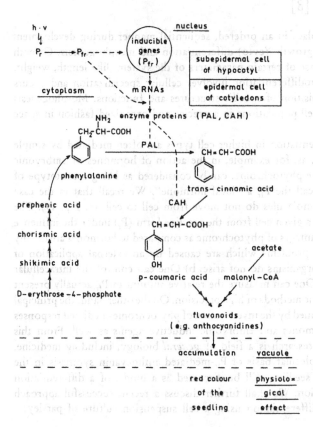

CYANIDIN

Fig. 158. The formula of the anthocyanidin cyanidin

Fig. 159. This scheme is intended to emphasize the significance of the enzyme phenylalanine ammonia-lyase (PAL) as key enzyme of the phenylpropanoid metabolism. The biosynthetic pathways which depart from trans-cinnamic acid eventually lead to anthocyanidins, among other secondary plant products such as lignin (cf. Fig. 76). The enzyme trans-cinnamic acid 4-hydroxylase (CAH) which catalyzes the formation of p-coumaric acid from trans-cinnamic acid, can also be induced by phytochrome [226] (modified after [176a])

Fig. 159 describes in principle the biochemical pathway by which cyanidin is formed, and it illustrates at the same time the hypothesis which explains the action of P_{fr} on cyanidin biosynthesis in terms of differential enzyme induction. The molecular skeleton of the flavonoids (including cyanidin) is derived from the cinnamic acid pathway as well as from the malonyl CoA pathway. The hypothesis assumes that the control by P_{fr} is exerted along the cinnamic acid pathway through "differential enzyme induction". The double function of P_{fr} in mediating anthocyanin synthesis in the mustard seedling (at the level of transcription as well as at the level of translation, cf. 12th lecture) is taken into account in the model. The key enzyme is phenylalanine ammonia-lyase (PAL), an enzyme which catalyzes the formation of transcinnamic acid from phenylalanine and thus connects the phenylpropanoid pathway of secondary metabolism to the basic metabolism.

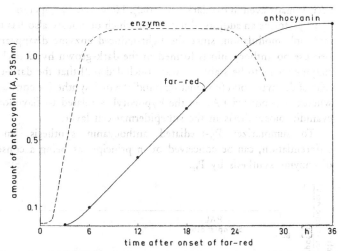

Fig. 160. A hypothetical illustration (broken line) of a situation in which P_{fr}, maintained in the mustard seedling by continuous far-red light, causes the synthesis of an enzyme which is rate-limiting during the linear phase ("production phase") of anthocyanin synthesis

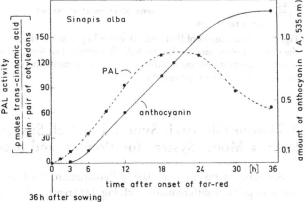

Fig. 161. Time-courses of anthocyanin contents and PAL activity in the cotyledons of the mustard seedling. Dark values do not deviate significantly from zero in both cases. While PAL activity increases up to 18 h after the onset of light, the *rate* of anthocyanin synthesis is constant between 6 and 24 h after the onset of light

The model in Fig. 159 has been supported over the years by many experimental data including inhibitor experiments (e.g. Fig. 71) and direct enzyme measurements (e.g. Fig. 77). We recall that in the dark-grown mustard seedling no significant amounts of PAL are detectable in the cotyledons. However, the enzyme is induced by P_{fr}, operationally, by continuous far-red light (P_{fr}^+) as well as by $P_{fr\ (ground\ state)}$ (cf. 4th lecture). We further recall that PAL is not stable but disappears through inactivation (cf. Fig. 80).

The problem was whether or not PAL could be considered to be the rate-limiting enzyme for cyanidin formation (Fig. 160). Fig. 161 clearly shows that this is not the case. Obviously there are other enzymes which are rate-limiting during the "production phase" of anthocyanin synthesis, among them CAH = trans-cinnamic acid Hydroxylase[226].

A second problem arises from the fact that in the hypocotyl of the mustard seedling

PAL is made even in the dark (Fig. 162). However, P_{fr} (operationally, continuous far-red light) performs an additional induction which obviously also has the character of a fully-reversible modulation, since the light-induced enzyme disappears through inactivation. Because no anthocyanin is formed in the dark-grown hypocotyl and because the dark-enzyme seems to be stable, it was concluded [49] that the dark-enzyme occurs in those cells of the hypocotyl (conducting bundle) which synthesize considerable amounts of lignin, whereas P_{fr}-induced PAL in the hypocotyl is related to flavonoid biosynthesis, e.g. to cyanidin biosynthesis in the subepidermal cell layer.

To summarize: P_{fr}-mediated anthocyanin synthesis, an event of secondary differentiation, can be conceived of in principle as being a consequence of modulation of enzyme synthesis by P_{fr}.

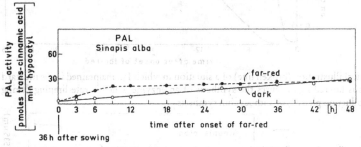

Fig. 162. Time-courses of PAL activity in the hypocotyl of the mustard seedling in the dark and under the influence of continuous far-red light. Remember that longitudinal growth of the hypocotyl occurs at nearly constant rates during the period of experimentation in the dark as well as under continuous far-red light (cf. Fig. 65) (after [49])

3. Flavone Glycoside Synthesis in Cell Suspension Cultures of Parsley as a Model System for Primary and Secondary Differentiation

In our experiments on secondary differentiation, the factors which determine the time-course of primary differentiation (cf. Fig. 148) remained unknown. We recall that the "factors of primary differentiation" determine the pattern of competence (with respect to P_{fr}) in the multicellular system. Recently E. WELLMANN used cell suspension cultures to develop a model system for the investigation of primary differentiation [294]. Since it has not been possible so far to obtain cell suspension cultures from mustard, WELLMANN used cell suspension cultures from parsley (*Petroselinum hortense*). These cultures were already known to form large amounts of flavone glycosides under high-irradiance white light [89].

Fresh weight was used in WELLMANN's experiments as a system of reference, since it was independent of the irradiation applied. However, total protein or a marker enzyme of the basic metabolism, e.g. glucose-6-phosphate dehydrogenase, could have been used instead with exactly the same result.

Preliminary "action spectra" for the light mediated induction of flavone glycosides in the parsley cell suspension culture were obtained with cut-off filters as well as with a monochromator (Table 15). The spectra indicate that visible light is ineffective, and that the peak of activity is below 300 nm, that is, in the range of maximum absorption by nucleic acids. The fluorescent white light which was used previously [89] was only effective

Table 15. Estimation of the spectral range active in flavone glycoside accumulation in cells of parsley by means of ultraviolet-absorbing cut-off filters. The samples were irradiated with white fluorescent light for 2 h and flavone glycosides measured after a further 20 h of darkness (after [294])

Filter Used $\lambda_H{}^a$ [nm]	Flavone Glycosides $A_{380}{}^b$
435	0.13
385	0.19
345	0.31
320	0.73
280	0.90
without	0.92
dark control	0.12

[a] Wavelength at which transmittance of the filter is 50%.
[b] Standard extract: 1 g fresh weight/5 ml buffer.

Table 16. Test for light-dependent changes in fresh weight, protein content and extractable activity of glucose-6-phosphate dehydrogenase in a cell suspension culture of parsley (after [294])

Program of Irradiation	Fresh Weight (g/sample)	Protein (mg/ml extract)	Glucose-6-phosphate dehydrogenase ($\Delta A/min$)
60 min UV + 10 min red + 15 h dark	1.41	0.85	0.119
60 min UV + 10 min far-red + 15 h dark	1.39	0.88	0.128
60 min UV + 15 h far-red	1.42	0.86	0.123
16 h dark	1.37	0.86	0.125

Table 17. Flavone glycoside accumulation mediated by short-time irradiation with red and far-red light and with continuous far-red light in a cell suspension culture of parsley (after [294])

Program	Flavone Glycosides^a (A_{380})
Preirradiation with 60 min ultraviolet followed by:	
15 h dark	0.36
15 h far-red	0.41
10 min red + 15 h dark	0.355
10 min red + 10 min far-red + 15 h dark	0.255
10 min far-red + 15 h dark	0.26
10 min far-red + 10 min red + 15 h dark	0.35
Without ultraviolet preirradiation:	
16 h red	0.12
16 h far-red	0.125
10 min red + 16 h dark	0.12
10 min far-red + 16 h dark	0.125
16 h dark	0.125
Initial value before irradiation	0.12

[a] Standard Extract: 1 g fresh weight/5 ml buffer.

because it contains a relatively high contamination with short-wavelength UV. During the period of experimentation fresh weight increased by the same amount in either dark-treated or irradiated cultures (Table 16). No differences in protein content due to the irradiation could be detected. Furthermore the activity of glucose-6-phosphate dehydrogenase, a marker enzyme of the basic metabolism, remained unchanged. There was no indication of any damage to the cells due to the standard ultraviolet (UV) used. A linear increase in flavone glycoside accumulation was measured with respect to the time of UV irradiation up to 5 h whereas in the actual experiments only 60 min of standard UV were used.

The main results are indicated on the upper part of Table 17. These data show that phytochrome can become effective in the cells only after preirradiation with UV. The stimulation of flavone glycoside accumulation caused by 60 min of standard UV was reduced by about 40 per cent by a subsequent pulse of 10 min of far-red light. This reduction of the UV effect was nullified by a subsequent red irradiation. Thus, the operational criteria for the involvement of phytochrome are clearly fulfilled. However, red and far-red light, given without ultraviolet pretreatment, had no effect. The fact that red light given after UV showed no additional stimulation of flavone glycoside accumulation, was to be expected. The UV by itself will lead to the formation of a large percentage of P_{fr}. Another indication for the involvement of phytochrome is the result that continuous far-red light will increase flavone glycoside accumulation to values considerably higher than those obtained after UV followed by darkness. It must be emphasized that there was no effect of continuous red or far-red light upon flavone glycoside formation without a preceding irradiation with UV light (lower part of Table 17).

The data so far obtained with the cell cultures suggest a model system for the study of the double-action control mechanism in differentiation. Obviously UV can be used to make the cells competent for P_{fr}. It can be suggested that UV exerts some specific action, which changes the cells in such a way that phytochrome becomes able to act on the differentiation process as measured by the synthesis of flavone glycosides or enzyme induction (Table 18).

4. Some General Remarks

We realize, of course, that biochemical model systems of differentiation which include only simple, water-soluble end-products (such as flavone glycosides, anthocyanins), will

Table 18. Control of PAL increase in cell suspension cultures of parsley by standard ultraviolet (= UV), red and far-red (= fr) light (after [295])

Program	PAL Activity (relative units)
20 min UV + 300 min dark	100
20 min UV + 5 min fr + 295 min dark	74 ± 2
20 min UV + 5 min fr + 5 min red + 290 min dark	99 ± 3
20 min UV + 5 min fr + 5 min red + 5 min fr + 285 min dark	73 ± 2
5 min fr + 315 min dark	8.5 ± 0.7
5 min fr + 5 min red + 310 min dark	8.4 ± 0.8
320 min dark	8.6 ± 0.7

probably not lead to a complete solution of the real problem, i.e. the relationship between enzymes and form, or, in other words, the relationship between enzyme specificity and structural specificity. However, we believe that studies of simple biochemical model systems of differentiation will increase our understanding of the logical principles and molecular foundations of differentiation and thus will eventually open a way to approach the basic problem of developmental biology, which is the relationship between modulation of enzyme synthesis and development of structural specificity, in space and time.

Appendix: Phytochrome-mediated Anthocyanin Synthesis as a Model System for Two-factor Analysis (Multiplicative Calculation, cf. 9th Lecture)

The system used is the anthocyanin-producing anabolic sequence in the competent cells of the mustard seedling; the two factors are "standard far-red light" (fr) and the antibiotic "chloramphenicol" (CAP, cf. Fig. 70).

Although CAP is a potent inhibitor of protein synthesis in bacteria, it usually has little effect on protein synthesis in non-bacterial systems. Higher plants constitute an outstanding exception insofar as CAP depresses the synthesis of chloroplast protein, the synthesis of cytoplasmic protein being relatively unaffected. CAP within a certain range of concentrations [20–40 μg · ml^{-1}] *increases* the rate of far-red-mediated anthocyanin accumulation in the mustard seedling. The initial lag-phase after the onset of far-red light and the time of termination of anthocyanin synthesis are not influenced by the presence of the antibiotic (Fig. 163). With and without CAP a constant rate of anthocyanin accumulation is observed over a period of at least 24 h after the initial lag-phase at all far-red irradiances investigated. It is found that the relative increase of the rate of anthocyanin accumulation which is due to CAP [20 μg · ml^{-1}] is independent of the far-red irradiance applied (Table 19). In total darkness CAP does not cause any anthocyanin synthesis. Formally one can conclude that CAP and far-red light act as two independent

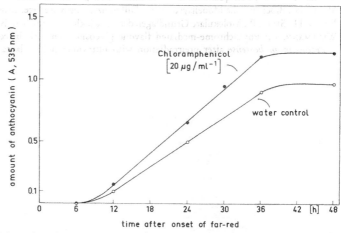

Fig. 163. The kinetics of far-red-mediated anthocyanin accumulation in the mustard seedling with and without chloramphenicol [20 μg · ml^{-1}]. Onset of light: 24 h after sowing (after [283])

factors in a multiplicative system (cf. Fig. 110). On the "molecular" level the observations can very probably be explained as follows: CAP at a concentration of [20 µg · ml^{-1}] inhibits the protein synthesis of the plastids. This inhibition leads to an increase of the precursor pool of phenylalanine in the cotyledons. Because the small concentration of CAP does not interfere with protein synthesis (enzyme synthesis) in the cytoplasm, far-red-mediated anthocyanin synthesis can proceed normally. Since phenylalanine acts as a precursor of flavonoids, the increased pool of this substance will lead to an increase in the rate of anthocyanin accumulation.

Table 19. Anthocyanin accumulation in the mustard seedling under the simultaneous influence of standard far-red light (fr) (that is, P_{fr}*) and chloramphenicol (CAP) [20 µg · ml^{-1}] as an example of a two-factor analysis. If "multiplicative calculation" is obeyed, the ratio CAP value/water value must be the same under all circumstances. Anthocyanin was measured 15 h after the onset of far-red light. Remember that the physiological effectiveness of standard far-red light is a function of irradiance over a wide range (cf. Fig. 52) (after [165])

Relative Irradiance	Amount of Anthocyanin with CAP (= CAP value)	Amount of Anthocyanin without CAP (= Water value)	$\dfrac{\text{CAP value}}{\text{H}_2\text{O value}}$
1/1 fr	0.293	0.198	1.48
1/10 fr	0.228	0.149	1.53
1/100 fr	0.125	0.081	1.54

Prediction (on the basis of multiplicative calculation): CAP value = constant × water value.
Experimental result: CAP value = 1.52 × water value.

Suggested Further Reading

HAHLBROCK, K., EBEL, J., ORTMANN, R., SUTTER, A., WELLMANN, E., GRISEBACH, H.: Regulation of enzyme activities related to the biosynthesis of flavone glycosides in cell suspension cultures of parsley (*Petroselinum hortense*). Biochim. biophys. Acta 244, 7 (1971).
MOHR, H.: Biologie als quantitative Wissenschaft. Beilage zu: Naturwiss. Rdsch., Heft 7, 779 (1970).
MOHR, H.: Biochemische Modellsysteme für Differenzierungsvorgänge. Umschau 1971, p. 547.
MOHR, H., SITTE, P.: Molekulare Grundlagen der Entwicklung. München: BLV 1971.
WELLMANN, E.: Phytochrome-mediated flavone glycoside synthesis in cell suspension cultures of *Petroselinum hortense* after preirradiation with ultraviolet light. Planta 101, 283 (1971).

Control of Distinct Enzymes (PAL, AO) in Different Organs of a Plant (Mustard Seedling)

1. The Problem

In microbial systems as well as in animals or plants we must deal with the problem of how changes are effected in the metabolic machinery, i.e. in specific enzymes, in response to environmental and nutritional conditions or as part of a developmental sequence. Such changes include removal of unneeded enzymes as well as the synthesis of those newly required. While in bacteria the removal process can involve dilution during phases of rapid growth, in animal [252] or plant tissue where little or no cell division takes place the process of protein degradation is essential as a means of removing unneeded metabolic machinery, and therefore as a means of controlling enzyme levels.

In plant tissue (cotyledons of the mustard seedling) two types of enzyme have been observed in connection with phytochrome-mediated enzyme induction during photomorphogenesis: 1) Enzymes (e. g. amylase, glycollate oxidase, ascorbate oxidase) which seem to be stable during the period of experimentation; and 2) Enzymes (e.g. phenylalanine ammonia-lyase = PAL), the activity of which eventually returns to a low level.

In the present lecture we will concentrate on the enzyme phenylalanine ammonia-lyase (PAL) from the following viewpoint: Is the behavior of this particular enzyme with respect to inducibility by phytochrome (P_{fr}) and with respect to disappearance (degradation) the same in different organs (cotyledons and hypocotyl) of a seedling (*Sinapis alba*), or is the behavior of the enzyme determined by the nature of the organ?

2. The Advantages of the Experimental System Used for these Investigations

The advantages of the experimental system used (mustard seedling) for the problem in question can be summarized as follows: a) In neither organ (cotyledons or hypocotyl) is there a significant increase in the amount of DNA or of cell number during the experimental period [288]. [Since in the taproot (=radicle) an increase in DNA and cell number must be anticipated during the experimental period, this organ was not included in the present investigation.] b) PAL from both organs (cotyledons, hypocotyl) appears only as a single band on gel electrophoresis [260]. There are no indications from SCHOPFER's studies that the enzyme is different in the two organs [260]. c) The increase in PAL activity is very probably due to an increase in the number of enzyme molecules. This problem has been checked repeatedly in the usual way, i.e. by demonstrating that drugs like Puromycin, Cycloheximide and Actinomycin D will inhibit the increase in enzyme activity [216], as well as by density-labelling with deuterium of the newly-formed enzyme [263]. d) The induction of PAL by light in the mustard seedling is very probably exclusively due to P_{fr}, the active species ("effector molecule") of the phytochrome system [95]. However,

P_{fr} can be active from the ground state (in the dark) as well as from some excited state (in continuous light) [264].

3. PAL: Results and Conclusions

In the dark-grown mustard cotyledons PAL activity can scarcely be detected by the assay used. However, the enzyme can rapidly be induced by continuous far-red light.

The problem was to explain the "basic kinetics", i.e. the time-course of PAL levels in the mustard cotyledons under the influence of *continuous* far-red light. The experimental basis of any explanation of the "basic kinetics" is found in the far-red → dark kinetics elaborated up to the peak of the "basic kinetics" (Fig. 164). The theoretical basis of these experiments (as far as phytochrome is concerned) is briefly the following: When the far-red light is turned off, the physiological effectiveness of the established P_{fr} concentration drops instantaneously to a low level, possibly to zero (cf. Fig. 135). The reason for this change is that the physiological effectiveness of a steady state P_{fr} concentration (maintained by standard far-red light) is a function of the quantum flux density over a wide range (cf. 4th lecture).

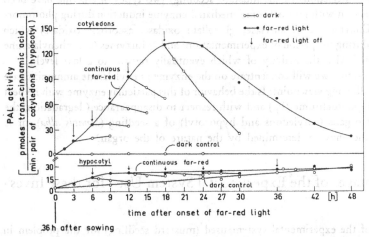

Fig. 164. The kinetics of the enzyme PAL in the cotyledons and in the hypocotyl of the mustard seedling in the dark and under the influence of continuous standard far-red light. Onset of light: 36 h after sowing. In addition, a number of far-red → dark kinetics are indicated. This term is used to designate those kinetics of the enzyme which are observed after the standard far-red light has been turned off (at arrows) (after [49])

Against this background Fig. 164 shows that P_{fr} is continuously required in order to maintain an increase of PAL. Further, the kinetics indicate that a degradation (irreversible disappearance) of PAL plays an important role even before the peak of the basic kinetics has been reached. It is preferable to interpret the basic kinetics of PAL in the mustard cotyledons by three principles (cf. Fig. 80): 1) Induction of PAL synthesis by P_{fr}, whereby P_{fr} is continuously required. 2) Increasing inactivation (degradation) of PAL by an unknown inactivating principle which comes into play at least 6 h after the onset of light. 3) Repression of PAL synthesis which comes into play at the peak of the basic kinetics. "Repression" in this context means that for some reason or other the PAL-producing system in the cotyledons no longer responds to P_{fr} with the formation of PAL.

In the hypocotyl (as well as in the taproot) PAL appears in the dark (Fig. 164). Furthermore, the basic kinetics of PAL, i.e. the time-course of PAL levels under continuous standard far-red light, are totally different as compared to the basic kinetics in the cotyledons. The preferable explanation of the hypocotyl data is based on the assumption that PAL synthesis in the dark and PAL synthesis mediated by phytochrome are totally independent phenomena. If the dark kinetics are subtracted from the far-red kinetics, a time-course of the "remainder enzyme" is found which is similar to the far-red kinetics in the cotyledons and possibly explicable in the same way. Since the "dark enzyme" and the "P_{fr}-dependent enzyme" do not differ in gel electrophoresis [260] or in any other aspect tested so far, the conclusion is justified that the two enzymes are identical. The situation in the hypocotyl can then be understood as follows: There are tissues in the hypocotyl (e.g. in the differentiating xylem [223, 224]) which produce the enzyme in the dark. This enzyme is stable, at least during the period of experimentation. On the other hand, there are tissues in the hypocotyl (e.g. the anthocyanin-synthesizing subepidermal layer) which produce the enzyme only under the control of P_{fr}, as is the case in the cotyledons. These two responses are independent of one another. If this interpretation is correct, the only difference between the cotyledons and the hypocotyl with respect to PAL would be that in the one organ there is considerable dark synthesis of a stable enzyme while in the other organ dark synthesis of PAL is scarcely detectable.

While it is possible to interpret the PAL data of Fig. 164 with more complicated models [303], the basic conclusion can hardly be circumvented. Although there is an obvious and seemingly complicated difference in the behavior of this particular enzyme in the two organs, it is highly probable that this difference is mainly (or even exclusively) due to the existence of a considerable dark synthesis of the enzyme in the hypocotyl. The factors controlling dark synthesis in the mustard seedling are totally unknown.

From these facts it is evident that the regulation of enzyme levels in organized plant tissues is complex, and that multiple mechanisms exist for controlling the level of a particular enzyme in a particular organ. Obviously there is no single or simple mechanism which controls the levels of an enzyme in all instances. Rather, as in the case of mammalian

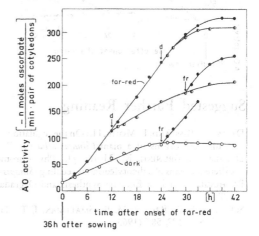

Fig. 165. The kinetics of the enzyme AO in the cotyledons of the mustard seedling in the dark and under the influence of standard far-red light (fr). In addition, two far-red → dark kinetics are indicated (d) (after [52])

tissues [252], we are led to the conclusion that control of the level of a particular enzyme may be exerted at any point at which control can be potentially exerted, and that this will depend upon the enzyme and the tissue involved.

4. Ascorbate Oxidase (AO): Results and Conclusions

In addition to PAL the behavior of the enzyme AO (soluble activity) has been compared in cotyledons and hypocotyl of the mustard seedling [52]. Fig. 165, 166 clearly indicate that the action of standard far-red light (that is, P_{fr}*) is precisely the same in both organs. However, the enzyme kinetics under the influence of far-red light ("basic kinetics") are conspicuously different in the two organs. While the dark kinetics look very similar, the light-induced enzyme seems to be considerably less stable in the hypocotyl than in the cotyledons at least up to 78 h after sowing. From about 18 h after the onset of light the constant AO level in the hypocotyl obviously represents a steady state (that is, the rate of far-red-mediated synthesis equals the rate of decay). On the other hand, there is no indication of an enzyme decay in the cotyledons during the experimental period. These results and the data on PAL as reported in the previous section lead to the following conclusion: The effect of phytochrome on enzyme induction is the same in both organs, cotyledons and hypocotyl. However, the processes of enzyme degradation are *specific* for the organ and for the enzyme. While in the case of PAL the phytochrome-induced enzyme behaves very similarly in cotyledons and hypocotyl, the far-red kinetics of AO are conspicuously different in the two organs. This difference can be attributed to a different decay of the enzyme in the two organs during the experimental period. Thus, the fine-structure of the time-courses of enzyme levels can be determined by the nature of the particular organ (that is, by the state of differentiation), even if the "mechanism" of the *inductive* process is precisely the same in the diferent organs.

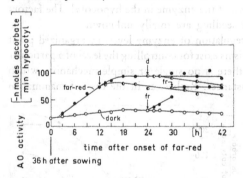

Fig. 166. The kinetics of the enzyme AO in the hypocotyl of the mustard seedling in the dark and under the influence of standard far-red light (fr). In addition, two far-red → dark kinetics are indicated (d) (after [52])

Suggested Further Reading

DITTES, L., RISSLAND, I., MOHR, H.: On the regulation of enzyme levels (phenylalanine ammonia-lyase) in different organs of a plant (*Sinapis alba* L.). Z. Naturforschg. *26b*, 1175 (1971).

DRUMM, II., BRÜNING, K., MOHR, II.: Phytochrome mediated induction of ascorbate oxidase in different organs of a dicotyledonous seedling (*Sinapis alba* L.). Planta *106*, 259 (1972).

RECHCIGL, M. (edit.): Enzyme Synthesis and Degradation in Mammalian Systems. Basel: S. Karger 1971.

SCHIMKE, R. T., BROWN, M. B., SMALLMAN, E.T.: Turnover of rat liver arginase. Ann. New York Acad. Sci. *102*, 587 (1963).

Energetics of Morphogenesis

This lecture is concerned with the energetics of a multicellular system during controlled morphogenesis. The system referred to is mainly the mustard seedling performing photomorphogenesis. The controlling factor is the active phytochrome P_{fr} (operationally, continuous far-red light, cf. 4th lecture).

1. The Background

1. The 16th lecture described how ascorbate oxidase (AO) can be induced by P_{fr} in the mustard seedling. It has been suggested that AO serves as a terminal oxidase of a respiratory chain in which the system ascorbate/dehydroascorbate functions as a biochemical redox system (Fig. 167). However, the electron flow in this respiratory chain does not seem to be related to ATP production; that is, the energy liberated along this chain can to our present knowledge not be preserved as free energy. Thus, the biological function of the postulated AO respiratory chain is not clear.

2. In a developing system, phytochrome is mainly concerned with anabolism, that is, with permitting the build-up and maintenance of complex metabolic and structural patterns.

3. FRIEDERICH [71] has shown in investigations with whole mustard seedlings that during the course of phytochrome-mediated photomorphogenesis the yield of plant material, as measured by the increase in fat-free dry matter divided by the decrease in fat, and the energy yield, as measured by the increase in heat of combustion of fat-free dry matter divided by the decrease in heat of combustion of fat, were reduced (by approximately 60 per cent). This result (decrease of energy yield in the course of photomorphogenesis) can be explained in two ways [71]. Firstly, assuming that in principle a change in the heat of combustion (ΔU) represents a corresponding change in Gibbs free energy (ΔG), it can be concluded that P_{fr} decreases the yield of Gibbs free energy. Such a result can be anticipated: under the influence of P_{fr} a great number of metabolic and anabolic processes take place which all together augment the deviation of the system from thermodynamic equilibrium, the latter being defined as $\Delta G = 0$. This in turn will increase the metabolic and morphogenic capacity of the system [108] but necessarily (i.e. for physical reasons) decrease the yield of Gibbs free energy. Secondly, if during photomorphogenesis the increase in Gibbs free energy (ΔG) of the fat-free dry matter is larger than the increase

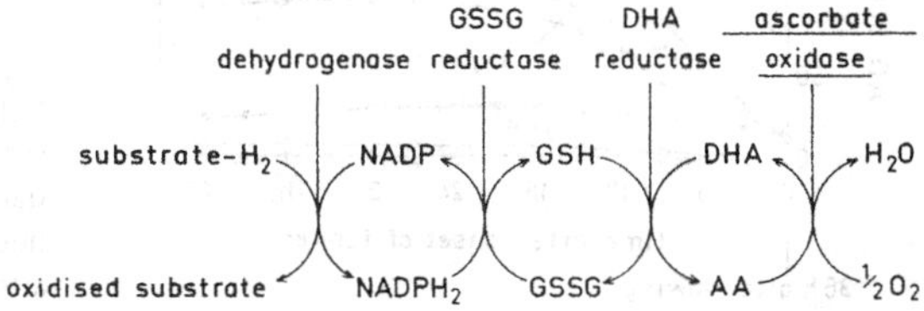

Fig. 167. A simple formulation for a respiratory chain with ascorbate oxidase as terminal oxidase. GSH, glutathione; AA, ascorbic acid (= ascorbate); DHA, dehydroascorbic acid (= dehydroascorbate) (after [3])

in heat of combustion (ΔU), the yield of Gibbs free energy could possibly be as high as in the dark (ΔG $\approx$ ΔU-TΔS). In this case the decrease of the energy yield (with respect to the heat of combustion) would mean that during photomorphogenesis those anabolic processes are favored which lead to substances and structures with a relatively low entropy content (formation of negative entropy within the seedling). However, since the seedling which develops under isothermal conditions with its environment only takes up water and oxygen, the total entropy of the seedling can hardly decrease. Therefore the second alternative seems improbable. In any case the scarcity of thermodynamic data for biologically important macromolecules (including structural proteins and even polysaccharides) has so far prevented a follow-up of this approach. Another approach to the problem has therefore been tried using the fact that AO can be induced by phytochrome in the mustard seedling. Unfortunately this matter is still largely speculative. However, it has already led to some conclusions with respect to "etiolation" and "regressive evolution" which deserve attention.

2. The Hypothesis

It was assumed that the operation of the AO respiratory chain in the mustard seedling contributes significantly to decreasing the energy yield (and correspondingly the yield of Gibbs free energy) and that the amount of AO is rate-determining in this pathway. It was further assumed that the phytochrome-mediated induction of *anabolic* processes (i.e. an increase in metabolic and structural complexity) is correlated with a decrease of the yield of Gibbs free energy. On the basis of this hypothesis it can be predicted that in the taproot where the morphogenic influence of P_{fr} is only small (cf. Fig. 1), the P_{fr}-mediated increase of AO will only be small whereas in the shoot system (hypocotyl as well as cotyledons) the increase of AO (as mediated by continuous far-red light) will be large. Fig. 168 shows that this prediction is verified. The fact that the relative increase of AO (as mediated by continuous far-red light) is virtually the same in hypocotyl and cotyledons

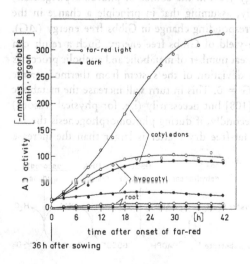

36 h after sowing

Fig. 168 Time-courses of ascorbate oxidase (AO) in the different organs of the mustard seedling (cotyledons, hypocotyl, taproot) in darkness and under the control of continuous far-red light. Onset of light: 36 h after sowing (after [52])

(and possibly in the taproot as well) is significant for the following reasons: the conspicuous photomorphogenic behavior of the two organs is totally different (e.g., inhibition of growth in the hypocotyl, promotion of growth in the cotyledons); other inducible enzymes (such as PAL [49] or amylase [53]) behave differently. In any case the far-red-mediated increase of AO cannot be related at present to any *specific* photomorphogenic response.

3. General Conclusions

We believe that the data obtained by Friederich [71] and the data on AO [52] bear on developmental physiology as well as on the theory of evolution. The argument is briefly the following: Any increase of metabolic capacity and structural complexity (increase of the deviation from thermodynamic equilibrium) will increase the flexibility and stability of the system towards external disturbances. However, at the same time the energy yield must necessarily (i.e. for physical reasons) decrease. Thus, a decreasing energy yield is the inevitable consequence of an ontogenetic or phylogenetic increase of metabolic and/or structural complexity. While in general the decreasing energy yield is tolerated for the sake of increasing capacity (in ontogeny as well as in phylogeny), it must be expected that a strong, negative selection pressure becomes effective as soon as a complex metabolic chain or a complex structure is no longer required by the system. Any decrease of complexity without a reduction of adaptation and flexibility under the prevailing conditions must be a selective advantage for the system.

4. Etiolation as an Adaptive Trait

Etiolation (cf. Fig. 1, 2) can be conceived of as a manifestation of a minimum deviation from thermodynamic equilibrium (maximum energy yield) at conditions of maximum lengthening of the system. On the other hand, photomorphogenesis is characterized by a rapid increase of structural and metabolic complexity and by a concomitant decrease of the energy yield. This can be tolerated by the plant since under natural conditions the activation of phytochrome is followed by the onset of photosynthesis. As far as the evolution of the angiosperms is concerned, it can be concluded that during etiolation of a seedling in complete darkness the selection pressure has been directed towards an increase of the energy yield whereas in the light (i.e. under conditions of photosynthesis) the selection pressure has been directed primarily towards increasing complexity and concomitantly increasing metabolic and structural flexibility at the cost of the energy yield.

5. The Rapidity of Regressive (or Degenerative) Evolution

A *negative* correlation between energy yield and complexity explains the rapid loss of complex characteristics (regressive evolution) as soon as these structures are no longer required by the system, e.g. chloroplasts in parasitic plants, eyes and pigmentation in cave-dwelling animals.

In bacteria (*Bacillus subtilis*, strain 168) it was shown that, on adequate media, "defective mutants", auxotrophs and others, indeed have a distinct selective advantage over the parental strain [302]. ZAMENHOF and EICHHORN [302] point out that under appropriate conditions even a mutant with a deletion would be at a distinct selective advantage over a point mutant and would supersede the latter. Such "mechanisms" could account for the complete disappearance of certain genes during the process of degenerative (or regressive) evolution.

A series of speculations have been made about the convergent reduction of eyes and pigmentation of cave animals [2]. While in general energetic aspects are neglected and the explanation of regressive evolution is based on "neutral" random mutations [296], BARR [2] interprets degenerative evolution of structures of cavernicolous animals as an indirect effect of pleiotropy: a mutation destructive for an eye is assumed to have a secondary pleiotropic effect which is positive for cave life. Whereas BARR expresses the opinion that selection under cave conditions acts by means of the pleiotropically positive effects of genes which have mutated degeneratively, it seems to be more appropriate to discuss this problem predominantly from the energetic aspect. In brief, the presumably scarce food supply in the caves is a strong selection factor which favors those degenerative mutations which increase the energy yield by decreasing the complexity of the system.

6. A Final Speculation

If it is generally true that the energy yield must drop with increasing complexity of metabolism and compartmentation it can be concluded that an upper limit for complexity exists in living systems which cannot be exceeded but only approached asymptotically. This would possibly make it understandable that a very high degree of metabolic complexity (such as secondary metabolism in angiosperms) and a very high degree of structural complexity (such as the central nervous system) could not have developed within the same living system.

Suggested Further Reading

BARR, T.C.: Cave ecology and the evolution of troglobites. In: Evolutionary Biology, Vol. 2. Amsterdam: North Holland Publishing Company 1968.

DRUMM, H., BRÜNING, K., MOHR, H.: Phytochrome-mediated induction of ascorbate oxidase in different organs of a dicotyledonous seedling (*Sinapis alba* L.). Planta (1972, in press).

FRIEDERICH, K.E.: Investigations on the energetics of phytochrome-mediated photomorphogenesis in mustard seedlings (*Sinapis alba* L.). Planta *84*, 81 (1969).

ZAMENHOF, S., EICHHORN, H.H.: Study of microbial evolution through loss of biosynthetic functions: establishment of "defective mutants". Nature (Lond.) *216*, 456 (1967).

Control of Plastogenesis by Phytochrome

1. The System: Cotyledons of the Mustard Seedling

The phytochrome-mediated responses which will be described in the course of this lecture take place in the cotyledons of the mustard seedling (Fig. 169). These cotyledons are peculiar organs. As long as the seedling develops in the dark exclusively, the cotyledons function as storage organs. Filled with storage fat and storage protein they serve the requirements of the rapidly-growing axis system, hypocotyl and taproot. The cotyledons do not grow or develop significantly as long as the seedling is kept in complete darkness. However, when the seedling is illuminated the cotyledons are transformed rapidly into photosynthetic organs, very similar in internal structure and in function to a normal photosynthetically

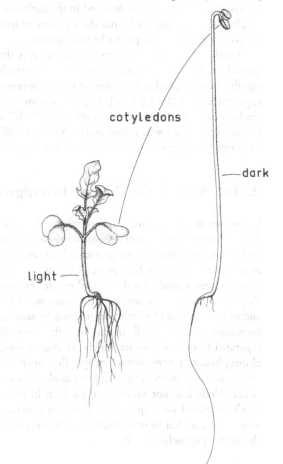

cotyledons

—dark

light —

Fig. 169. These two mustard seedlings (*Sinapis alba*) have the same chronological age and are virtually identical genetically. The differences in morphogenesis are due to light. The drawings emphasize the fact that light causes the cotyledons to develop from storage organs (right) to photosynthetically active leaves (left). Unlike for instance castor bean endosperm, the mustard cotyledon is not devoid of function after depletion of lipid and protein reserves, since in the light it expands, becomes green and persists as a photosynthetic organ

active leaf. The "mechanism" of the phototransformation of the cotyledons has been studied over the years, that is, the transformation by light of a storage organ into a photosynthetically active leaf.

Fig. 1 indicates that the morphogenic influence of light on cotyledon transformation is not due to photosynthesis. The effect of continuous standard far-red light which hardly allows any chlorophyll formation (cf. Fig. 60) is at least as strong as the morphogenic influence of the high-irradiance white light which does allow chlorophyll formation and photosynthesis. Indeed, the available information clearly shows that the *morphogenic* control exerted by light over cotyledon development is exclusively due to the formation of P_{fr}, whereby $P_{fr \text{ (ground state)}}$ as well as P_{fr}^* are effective (cf. 4th lecture).

The state of the phytochrome system in the mustard cotyledons has already been described (cf. Fig. 27). Let us just recall here that continuous standard far-red light maintains in the cotyledons of the mustard seedling something like a steady state of the phytochrome system with a relatively low but nearly stationary concentration of the active phytochrome, P_{fr}, over a considerable period of time. Irreversible decay of P_{fr} and *de novo* synthesis of P_r just compensate to yield an approximately photosteady state within the period of time used for the experimentation (as a rule from 36 to 72 h after sowing). The characteristic P_{fr}/P_{tot} ratio is established rapidly, i.e. in a matter of a minute, after the onset of standard far-red light. No P_{fr} can be detected in the dark-grown seedling. The only detectable form of phytochrome present in the dark-grown mustard seedling is P_r, the physiologically ineffective form of the phytochrome system.

Another important point to bear in mind is that between 36 and 72 h after sowing, growth of the mustard cotyledons is predominantly due to cellular growth. There is no significant increase of cell number or DNA contents in the cotyledons during the period of experimentation (25° C) [288]. For this reason the biological unit "pair of cotyledons" can be used as a system of reference instead of "cell" or "unit DNA". It must be emphasized that all experiments were performed with intact seedlings. The cotyledons were only dissected immediately prior to the cytological or biochemical analysis.

2. The Aim of the Present Investigations

In the past it has been usual to investigate the so-called "greening process" under white light and to follow the course of biosynthesis of plastid components and plastid structures under these conditions. This approach has been necessarily a purely descriptive one, since an *analysis* of the controlling processes requires a far-reaching separation of photomorphogenic processes, mediated through phytochrome, from secondary influences of photosynthesis. For this reason attempts are being made to analyze the control of plastogenesis and of biosynthesis of plastid components by standard far-red light without a considerable formation of chlorophyll. In this way the controlling function of phytochrome can be separated from those events which are consequences of the photochemical formation of chlorophyllide a from protochlorophyllide or of photosynthesis proper. The results show clearly that phytochrome probably controls every source of biosynthesis of plastid components. While it is not understood (at least in physiological terms) why in the etiolating seedling plastid development in darkness proceeds only to an etioplast stage and then stops, the conclusion is justified that the *resumption* of etioplast development is due to the active phytochrome, P_{fr}.

In the following some representative examples of phytochrome-mediated processes which occur in the cotyledons of the mustard seedling will be described. We will return to the problem of control of chlorophyll synthesis towards the end of this lecture.

3. Some Histological Data

In complete darkness, the histological structure of the cotyledons does not change significantly with time. However, irradiation with white light as well as with continuous far-red light will lead to conspicuous structural changes in the direction of a photosynthetic leaf. The appearance of intercellular space can easily be seen. A closer inspection reveals the appearance of spongy and palisade parenchyma [87].

Fig. 170 (left) shows part of the lower epidermis of a dark-grown mustard cotyledon. The cell walls are smooth and the stomata are not fully developed. Under the influence of standard far-red light the stomata rapidly mature and the epidermal cells expand and develop a wall structure characteristic of a mature epidermis (Fig. 170, right).

In connection with Fig. 74 the increase of cotyledon area was described in quantitative terms. We recall that the cotyledon area increases only slightly in darkness between 36 and 60 h after sowing whereas continuous far-red light induces a strong enlargement. Although the kinetics of this response are complicated, the important feature, that is, a *slight* enlargement of the cotyledons in darkness, and a strong and qualitatively different enlargement under the influence of continuous far-red light, can easily be recognized. It has already been shown that Actinomycin D [$10\mu g \cdot ml^{-1}$] will not influence the enlargement of the cotyledons in darkness whereas the antibiotic cancels the P_{fr}-mediated enlargement completely if it was applied before or at the onset of light (cf. Table 4). These data support the view that the action of P_{fr} on cotyledon expansion depends on RNA synthesis whereas dark expansion is not limited by the supply of RNA.

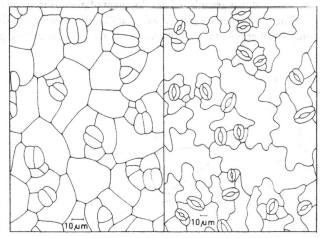

Fig. 170. Left, representative view of the lower epidermis of the cotyledons in dark-grown mustard seedlings; right, representative view of the lower epidermis of the cotyledons in light-grown mustard seedlings. Expansion of the epidermal cells and maturation of stomata are controlled by phytochrome; the involvement of $P_{fr \text{ (ground state)}}$ and of the "high irradiance response" have been demonstrated (after [137])

4. Control of Plastid Development by Phytochrome

Fig. 171 shows the kinetics of protein levels in the attached cotyledons of the mustard seedling in darkness and under the influence of standard far-red light. In order to understand these two curves one must realize that the cotyledons of the mustard seedling contain much storage protein which will be degraded during the development of the seedling. The resulting soluble compounds are either translocated to other parts of the seedling or used within the cotyledons for the synthesis of enzymes and structural proteins. This accounts for the strong *relative* increase in protein under the influence of far-red light. Notice that the seedling is a closed system for nitrogen over the whole period of experimentation. Therefore total nitrogen per seedling, the system of reference, is a constant.

Histochemical studies have shown that under the influence of far-red light, the degradation of storage protein in the cotyledons is enhanced. This fact is indicated in the central part and on the right-hand side of Fig. 172. It shows mesophyll cells of mustard cotyledons. The dense, dark bodies represent storage protein 72 h after sowing. It is evident, and it has been measured quantitatively [87], that far-red light enhances the degradation of the storage protein. At the same time, however, P_{fr} (operationally, continuous far-red light) stimulates a strong *de novo* synthesis of "structural protein" (in terms of histochemistry) in the cotyledons. This fact is indicated on the right-hand side of Fig. 172. The lens-shaped large, proteinaceous bodies which appear under the influence of far-red light have been called "far-red plastids". As far as number, size and shape are concerned, they are indistinguishable from chloroplasts under the light microscope except that they do not contain significant amounts of chlorophyll. The longest diameter of the plastids is about 5 μm, whereas that of the corresponding etioplasts is only about 1.5 μm. (The small etioplasts of the dark-grown seedling were not drawn). "Normal" green chloroplasts formed under white light in the cotyledons are shown on the left-hand side of Fig. 172. As already pointed out they do not differ in number, size and shape from the "far-red plastids". It is concluded that growth and development of plastids is under the control of phytochrome.

The electron microscope reveals that 72 h after sowing, the dark-grown seedling contains only small etioplasts, each of which has a somewhat diffuse prolamellar body and a few

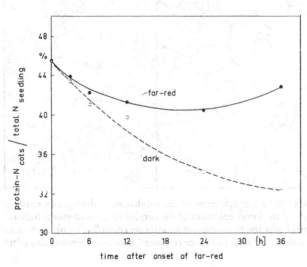

Fig. 171. Changes of protein contents in attached cotyledons of the mustard seedling in darkness and under continuous standard far-red light. (Remember that the amount of DNA in the cotyledons does not change significantly during the period of experimentation. Likewise, total nitrogen per seedling does not change) (after [120])

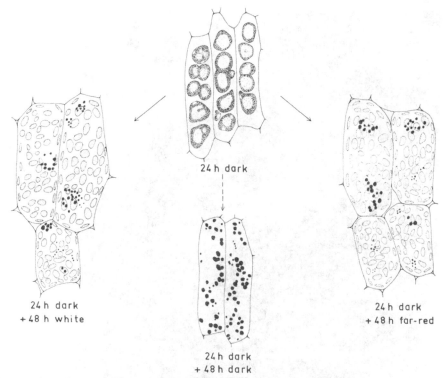

Fig. 172. These drawings illustrate the control exerted by continuous standard far-red light over degradation of storage protein (represented by dark, dense bodies 72 h after sowing) and formation of plastids in mesophyll cells of mustard seedling cotyledons. The small etioplasts of the dark-grown seedling were not drawn. Onset of light: 24 h after sowing (after [87])

short membranes (Fig. 173). Treatment with white light instead of darkness leads to the formation of normal cotyledon chloroplasts (Fig. 174). Treatment with far-red light leads to the formation of plastids which have the same size as those in white light but which possess an internal structure similar to etioplasts (Fig. 175). The structural substance is arranged as a very regular prolamellar body from which emerge a considerable number of thylakoids. The prolamellar body of the far-red plastids is a very conspicuous and regularly-organized feature. Fluorescent microscopy studies indicate that each "far-red plastid" has at least one localized center of fluorescence; this suggests that the prolamellar body is the site of protochlorophyllide deposition [87].

5. Control of Carotenoid Synthesis by Phytochrome

Carotenoids are typical constituents of the plastids. Since these terpenoids are very probably restricted to the plastid compartment and can be synthesized even in complete darkness, it can be assumed that the carotenoids (and not the chlorophylls) represent the plastid compartment as far as control by phytochrome is concerned. In brief, carotenoids can possibly serve as molecular representatives of the plastid compartment while the chlorophylls cannot serve this purpose.

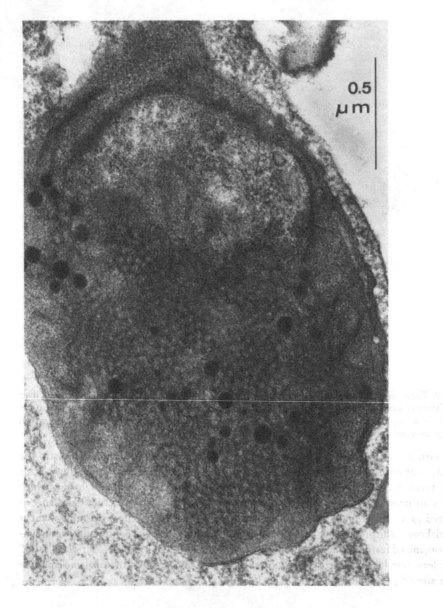

Fig. 173. Section through a typical etioplast from the cotyledonary mesophyll of a dark-grown mustard seedling, 72 h after sowing. Fixation: 3% glutaraldehyde, postosmicated (photographs taken by M. HÄCKER and H. FALK)

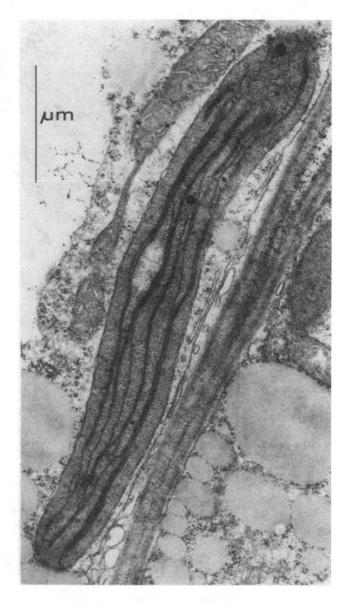

Fig. 174. Longitudinal section through a typical chloroplast from the cotyledonary mesophyll of a mustard seedling, 72 h after sowing. Treatment after sowing: 24 h dark, 48 h white light. Fixation: 3% glutaraldehyde, postosmicated (photographs taken by M. HÄCKER and H. FALK)

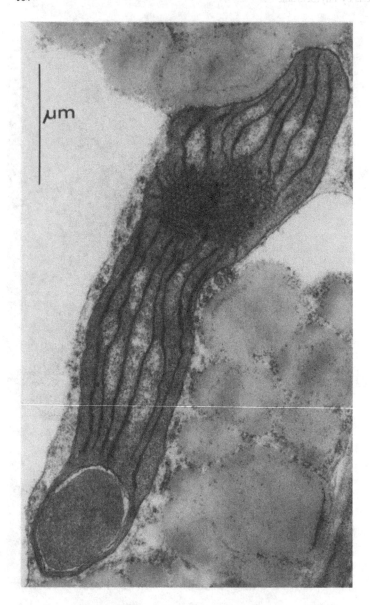

Fig. 175. Longitudinal section through a typical "far-red plastid" from the cotyledonary mesophyll of a mustard seedling, 72 h after sowing. Treatment after sowing: 24 h dark, 48 h standard far-red light. Fixation: 3% glutaraldehyde, postosmicated (photographs taken by M. HÄCKER and H. FALK)

This point is by no means a trivial one. GOODWIN, a world authority in the field of carotenoid biochemistry, has only recently expressed his opinion on this subject as follows: "As the light stimulated synthesis of the various terpenoids parallels the synthesis of chlorophyll, which can be taken as an indication of chloroplast development, they are obviously being synthesized alongside chlorophyll in the developing chloroplast ..." [81]. Fig. 99 and Table 10 remind us that GOODWIN is wrong: carotenoid synthesis in the mustard cotyledons responds to P_{fr} in a similar way as does the structural unit "plastid". Phytochrome-mediated plastid formation and carotenoid synthesis are correlated. Neither phenomenon is correlated with chlorophyll synthesis.

Table 10 not only proves that the operational criteria for the involvement of phytochrome (P_{fr} in the ground state) are fulfilled; in addition, the information in this table shows that the formation of chlorophyll a (or the lowering of the photoconvertible protochlorophyllide pool) is not related to the light-mediated response "increase of the rate of carotenoid synthesis". Repeated irradiation with 5 min of red light each time, results in the formation of a substantial amount of chlorophyll a. Since the protochlorophyllide → chlorophyllide a phototransformation is irreversible, the amount of chlorophyll a formed by the programs: 2 × (5 min red) and 2 × (5 min red + 5 min far-red) is virtually the same. However, the response (as measured by the increase of the rate of carotenoid synthesis) is very different under the two programs and can only be related to the action of the phytochrome system.

From the 8th lecture we recall that the control of carotenoid synthesis by phytochrome (operationally, by continuous far-red light) can be conceived of as modulation of a metabolic sequence by P_{fr}. There is a continuous requirement for P_{fr} (that is, rapid modulation of the rate of synthesis if the state of the phytochrome system changes); there is no detectable secondary lag-phase; the sensitivity towards Actinomycin D is low while the sensitivity towards Puromycin and Cycloheximide is high (Fig. 176). Since the inhibitor-dependent rate of carotenoid synthesis is established about 1.5 h after the application of the inhibitor, one may conclude that the half-life of the rate-limiting enzyme(s) of carotenoid synthesis is relatively short. Furthermore, the data indicate that undisturbed protein synthesis *in the cytoplasm* is required to maintain carotenoid biosynthesis in the plastid compartment.

At this point in the discussion I would like to make some statements about the use of inhibitors in investigations concerned with protein synthesis in the cytoplasm and in the plastids (or mitochondria).

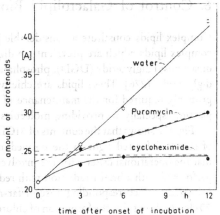

Fig. 176. Kinetics of carotenoid accumulation in the mustard seedling under continuous standard far-red light after the application of Cycloheximide [6 μg · ml⁻¹] or Puromycin [200 μg · ml⁻¹]. Onset of incubation with the antibiotics: 48 h after sowing. Onset of far-red light: 36 h after sowing (after [267])

Chloroplasts contain ribosomes (70s) which are distinct from those found in the cytoplasm (80s), but there has been no convincing identification so far of any of the proteins which the 70s ribosomes presumably synthesize.

However, chloroplasts are clearly not genetically autonomous; for genetic studies of both higher plants and algae indicate that at least some chloroplast components are encoded in the nuclear DNA [136].

In previous studies greening cells were treated with Chloramphenicol and Cycloheximide which selectively inhibit protein synthesis on ribosomes from chloroplasts and the cytoplasm, respectively. Unfortunately, different workers obtained conflicting results for some enzymes in both algae and higher plants. This may be due to the lack of *in vivo* specificity of the two antibiotics, especially Chloramphenicol [57].

Recently the antibiotic Lincomycin has been introduced as a chloroplast probe [57]. Lincomycin seems to be a very specific and highly effective inhibitor of the function of chloroplast ribosomes (70s). The main findings of the studies using this antibiotic are that chloroplast ribosomes are necessary for the synthesis of ribulose-diphosphate carboxylase and at least some of the chloroplast ribosomal proteins, but not for the synthesis of other enzymes in photosynthesis.

Returning to Fig. 176 it can be concluded that undisturbed protein synthesis in the cytoplasm is required to maintain carotenoid biosynthesis in the plastid compartment. From such results it must be inferred that specific mechanisms must exist for transporting proteins made on cytoplasmic ribosomes across the outer membranes of the proplastid (or etioplast). Even though the plastids may contain all the enzymes which are required to mediate carotenoid synthesis beyond the C_{15}-stage [82], the possibility must be considered that at least some (possibly most) of these enzymes were originally synthesized in the cytoplasm.

The rapid development of plastids under the control of standard far-red light (cf. Fig. 172) is evidence that the plastids use small molecules provided by heterotrophic metabolism from the extraplastidal cytoplasm. The required metabolites are supplied from the storage material in the mesophyll cells (cf. Fig. 172). A strong nutritional dependence of the developing plastid on the rest of the cell is therefore implied. Further, data such as those in Fig. 176 strongly indicate that the heterotrophy of the plastid is not restricted to metabolites but may also include enzymes.

6. Control of Galactolipid Biosynthesis by Phytochrome

Complex lipids constitute a considerable portion of the substance of plastids. The four complex lipids which are prevalent are diglycerides: monogalactosyl diglyceride (MGD), digalactosyl diglyceride (DGD), phosphatidyl glycerol and sulpholipid (sulphoquinovosyl diglyceride) [276]. These lipids are chiefly constituents of the plastid lamellae. They are probably required for the maintenance of the highly-organized structures present in the plastid, especially in providing molecules of a required geometry.

Fig. 177 shows that the contents of MGD and DGD increase strongly under the influence of standard far-red light in the cotyledons of the mustard seedling. Table 20 indicates that the operational criteria for the involvement of $P_{fr\ (ground\ state)}$ are fulfilled: an induction performed with a brief irradiation with red light is reversed if the red light is immediately followed by a corresponding dose of far-red light. In addition, the information in Table 20 clearly shows that the formation of chlorophyll a (or the lowering of the photoconvertible

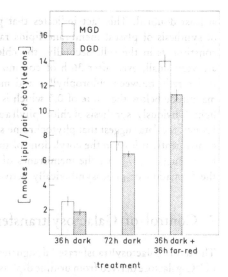

Fig. 177. The contents of monogalactosyl diglycerides (MGD) and digalactosyl diglycerides (DGD)
in the cotyledons of the mustard seedling kept
under standard conditions [162]. Treatment after
sowing: a, 36 h dark; b, 72 h dark; c, 36 h dark +
36 h standard far-red light (after [278])

protochlorophyllide pool) is not involved in the light-mediated responses. This point has
already been mentioned in connection with carotenoid synthesis, but it may be repeated
here since it has often been overlooked.

Repeated irradiation with 5 min of red light each time, results in the formation of
a substantial amount of chlorophyll a. Since the protochlorophyllide → chlorophyllide
a phototransformation is irreversible, the amount of chlorophyll a formed by the programs:
6 × red and 6 × [red + far-red] is similar and much larger than the tiny amount of chlorophyll formed by 6 × far-red. However, the response (as measured by the increase of galactolipids) is virtually the same with the programs 6 × far-red and 6 × [red + far-red] and
can thus only be related to the action of the phytochrome system.

Returning to Fig. 177, two points should be emphasized: Firstly, while the increase
of the total lipid-phosphate content in seedlings treated with standard far-red light is only
1.3 times that of seedlings grown for 72 h in the dark, the increase of galactolipids is

Table 20. Results of induction reversion experiments with the mustard seedling. Measurements of
the galactolipids were made 72 h after sowing. Standard far-red light [350 μW · cm^{-2}] and
standard red light [67.5 μW · cm^{-2}] were given for 5 min each at 36, 42, 48, 54, 60 and 66 h after
sowing (25 °C) (after [278]). Remember that the galactolipids are chiefly constituents of the plastid
"lamellae" [276]

Lipid	Treatment	$\left[\dfrac{\text{n moles lipid}}{\text{pair of cotyledons}}\right]$
MGD[a]	6 × red	14.58 ± 1.08
	6 × far-red	9.68 ± 0.80
	6 × (red + far-red)	10.68 ± 0.66
DGD[a]	6 × red	10.58 ± 0.46
	6 × far-red	6.74 ± 1.36
	6 × (red + far-red)	6.58 ± 0.44

[a] MGD = monogalactosyl diglyceride
DGD = digalactosyl diglyceride

at least doubled. This fact indicates that phytochrome leads to a specific enhancement of synthesis of plastid membrane lipids rather than to a general increase of membrane constituents in the cell. Secondly, the chlorophyll contents of far-red-treated seedlings are very small, even after 36 h of continuous far-red light (cf. Fig. 184). Therefore the molar ratio between chlorophyll a and monogalactosyl diglyceride is several orders of magnitude below the value of 0.5 which is usually reported in plants grown under white light. Obviously, synthesis of chlorophyll a and synthesis of galactolipids are not correlated. Rather, the data suggest that phytochrome specifically controls the biogenesis of thylakoid constituents, at least in the cotyledons of the mustard seedling. Lipid analyses of greening leaves suggest [17] that the membranes of the etioplast are used as building blocks for the formation of photosynthetically active thylakoids.

7. Control of Galactosyltransferase by Phytochrome

The term "galactosyltransferase" designates an enzyme which catalyzes the incorporation of ^{14}C-galactose derived from uridine diphosphate-galactose (UDP-galactose) into monogalactosyl and digalactosyl diglycerides [195]. The enzyme activity which can readily be isolated from the mustard cotyledons may be designated as monogalactosyltransferase since it transfers the radioactivity from UDP-galactose-^{14}C to an endogenous acceptor resulting predominantly in the formation of MGD. 90 per cent of the incorporated radioactivity is found in the monogalactosyl diglycerides and about 3 per cent in the digalactosyl diglycerides [277]. Recent findings by MUDD and associates [178] further support the conclusion that at least the first two of the three possible galactosylation steps are catalyzed by separate enzymes, in higher plants as well as in *Euglena*.

The kinetic results in Fig. 178 indicate that standard far-red light increases the amount of monogalactosyltransferase in the cotyledons of the mustard seedling. The enzyme activity reaches almost constant levels in the dark and under continuous far-red light at about 18 h after the onset of irradiation. In addition it was found that between 18 and 36 h after the onset of far-red light the accumulation of the galactolipids becomes linear at different rates in the dark and under far-red light [277]. Since the ratio of the enzyme

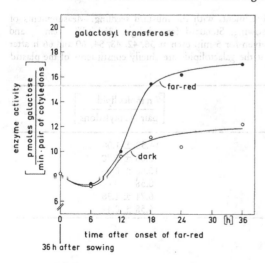

Fig. 178. Time-course of monogalactosyl transferase activity in the cotyledons of the mustard seedling in the dark and under continuous standard far-red light. Onset of light: 36 h after sowing (after [277]). The differences of the enzyme activities in the dark and under far-red light at 18, 24 and 36 h are statistically significant (paired-sample t-test)

activities is similar to the ratio of the accumulation rates of the galactolipids, the conclusion seems to be justified that the monogalactosyltransferase is the rate-limiting enzyme of galactolipid biosynthesis, at least during the period between 18 and 36 h after the onset of light.

8. Control of Glyceraldehyde-3-phosphate Dehydrogenase (GPD) by Phytochrome

The time-course of the NAD- as well as the NADP-linked enzyme was followed in the mustard cotyledons [36]. Both enzymes were measured in the same extract. The main finding was that the regulation of the two enzyme activities by standard far-red light is very different (Fig. 179). While the NAD-linked enzyme can always be detected, even in the dry seed, the NADP-linked enzyme becomes detectable only at 36 h after sowing. Furthermore, the *relative* induction due to standard far-red light, is much stronger in the case of NADP-linked GPD than with the NAD-linked enzyme. If the onset of light (standard far-red light) is only at 60 h after sowing, the NADP-linked enzyme is still induced to some extent while the NAD-dependent enzyme is not. These results seem to be a significant contribution to the question of whether or not the two enzyme activities are due to the same or to different proteins [23]. The answer from these data is that there are two distinct enzymes which are regulated differently in every respect.

The NADP-linked GPD which is localized in the plastids possibly represents the behavior of other enzymes of the photosynthetic carbon reduction cycle (Calvin cycle) as far as the regulation of synthesis by phytochrome is concerned [66]: there is some enzyme in the dark; P_{fr} exerts a strong induction; the enzyme is relatively stable over the whole period of experimentation. We recall (cf. Fig. 86) that the continued presence of P_{fr} is

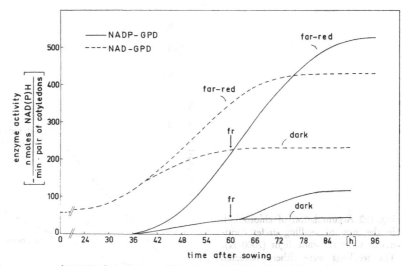

Fig. 179. Time-courses of NADP-linked glyceraldehyde-3-phosphate dehydrogenase (NADP-GPD) and NAD-linked glyceraldehyde-3-phosphate dehydrogenase (NAD-GPD) in the cotyledons of the mustard seedling in the dark and under continuous standard far-red light. Onset of light: at time of sowing or 60 h after sowing (after [36])

required to maintain GPD synthesis at a high rate. When the far-red light is turned off, the rate of enzyme synthesis decreases and eventually drops to zero.

The previously-mentioned antibiotic Lincomycin has been used recently to investigate further the question of where the NADP-dependent glyceraldehyde-3-phosphate dehydrogenase is actually synthesized [57]. While the synthesis of chloroplast ribosomal protein and the increase in ribulose-diphosphate carboxylase on illumination is almost totally inhibited by Lincomycin at [1 μg · ml⁻¹], the increase in NADP-linked GPD and DNA-dependent RNA polymerase is not affected, at least in illuminated pea apices. The present consensus among the workers in the field seems to be that the NADP-dependent glyceraldehyde-3-phosphate dehydrogenase is encoded in the nucleus, synthesized in the cytoplasm and translocated by a specific mechanism to the plastid compartment.

9. Control of Chlorophyll Synthesis by Phytochrome

a) Basic Phenomena and the Leading Model at Present

In Fig. 180 two populations of mustard seedlings are compared, one of which was pretreated with 24 h of standard far-red light while the other was kept in complete darkness. At 60 h after sowing both populations were placed under high irradiance white light (7000 lx) and chlorophyll accumulation was followed.

Greening of an etiolated leaf under constant illumination with white light is usually considered to occur in three stages (Fig. 180, lower curve): (1) The photoconversion of accumulated protochlorophyllide to chlorophyllide a. Since the photoconvertible protochlorophyllide pool is small in the mustard seedling, this step contributes only a small amount to the chlorophyll content of the green cotyledons. (2) The second stage is a lag-phase during which only a little additional chlorophyll accumulates. This lag-phase is two

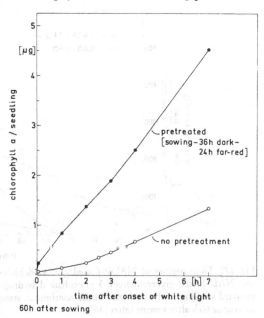

Fig. 180. Accumulation of chlorophyll in the mustard seedling under continuous standard white light (7000 lx). The seedlings were either pretreated with standard far-red light between 36 and 60 h after sowing (upper curve) or kept in complete darkness until the beginning of the white light (after [189])

hours in our material. (3) The third stage is a period of rapid chlorophyll synthesis which continues until the pigment content approaches that of the normal green cotyledon. The rate-limiting step during rapid chlorophyll synthesis when the light-dependent step (from protochlorophyllide to chlorophyllide a) is no longer rate-limiting, is not known. This can be achieved by using high irradiance white light (7000 lx in the present example).

Comparison of the two kinetics in Fig. 180 leads to the following conclusions:

1. The protochlorophyllide pool is somewhat (but not dramatically) increased by the pretreatment with P_{fr} (operationally, 24 h of continuous standard far-red light). The first points after the onset of white light were measured after 5 min of white light at 7000 lx illuminance. It has been shown that the total photoconvertible protochlorophyllide pool is indeed transformed at the end of the 5 min white light treatment.

2. The lag-phase is totally eliminated by the previous action of continuous far-red light.[1]

3. The rate of chlorophyll synthesis under steady state conditions is strongly increased by the pretreatment with continuous far-red light.

Two questions arise: Firstly, what is the molecular basis of phytochrome action (operationally, action of standard far-red light)? Secondly, what are the limiting factors during steady state accumulation of chlorophyll? These questions cannot be answered at the moment. NADLER and GRANICK [184] have proposed a tentative model (Fig. 181) which is

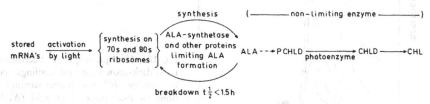

Fig. 181. A tentative working model for the control of chlorophyll synthesis in barley leaves. There is evidence that all the enzymes that convert δ-amino-levulinic acid (ALA) to chlorophyll are present in the plastids of higher plants in non-limiting amounts (after [184]). Recently, REBEIZ and CASTEL-FRANCO [215] have shown that isolated etioplasts are capable of synthesizing labelled protochlorophylls from ^{14}C-δ-ALA. Moreover, the light-stimulated incorporation of label from ^{14}C-labelled glycine and succinic acid into chlorophyll by isolated intact etioplasts [293] indicates the presence of δ-amino-levulinic acid (δ-ALA)-synthetase within the higher plant plastid

mainly based on inhibitor experiments and on results obtained by feeding δ-amino-levulinic acid (δ-ALA). The model proposes a light-mediated activation at the translational level of the synthesis of δ-ALA synthetase and other short-half-life proteins limiting δ-ALA formation. In our opinion this model does not explain the data in Fig. 180. While it would be premature to present an alternative model, some recent results obtained with the mustard seedling will be described briefly.

b) Feeding δ-ALA to Mustard Seedlings

The protochlorophyll(ide) in the etioplasts exists in at least three different forms, which can be characterized in situ by their absorption maxima in the red region: two of these forms, with absorption maxima at 650 nm and 636 nm respectively, are converted by light

---

1 The lag-phase can also be eliminated by brief preirradiations of the mustard seedling with red light. The effect of the red pulses can be fully reversed (in accordance with the operational criteria) by immediately following with a corresponding far-red pulse (KASEMIR et al., personal communication).

to chlorophyll(ide), while the third form with a peak at 628 nm is inactive (as far as the photoconversion to chlorophyllide is concerned) [1, 125].

Quantitative estimates of *isolated* protochlorophyll(ide) and chlorophyll(ide) are usually made from 80 per cent acetone extracts. However, the purity of the extract can be improved by transferring the pigments from acetone into ether ("ether extract").

The 624 nm peak of an ether extract of dark-grown mustard seedlings incubated on ALA was increased up to 15-fold over the peak of an extract from a non-treated control (Fig. 182). The 624 nm peak is attributed to protochlorophyll(ide) and the phenomenon

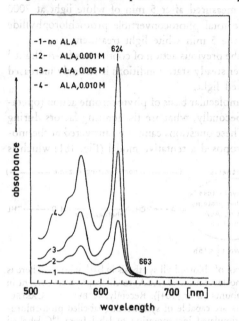

Fig. 182. Absorption spectra of acetone extracts from dark-grown mustard seedlings, incubated for one hour (47 to 48 h after sowing) in solutions of δ-amino-levulinic acid (ALA). The technique of incubation was described previously [143]. For determination of the spectra (at 60 h after sowing) the pigments were transferred into ether (after [129])

is interpreted as an increase of the protochlorophyll pool due to a non-limiting supply of the precursor ALA. However, the rate of chlorophyll(ide) formation (absorption peak at 663 nm) is not a function of the increasing protochlorophyll(ide) pool (Fig. 183). At least under standard far-red light which allows chlorophyll formation only at a very low rate, the rate of chlorophyll formation has no positive relation at all to the amount of material absorbing at 624 nm. Under white light of medium or high irradiance, chlorophyll (operationally, material absorbing at 663 nm) is rapidly destroyed in the presence of ALA. As a result, the application of ALA to the seedling leads to photodestruction of the chlorophyll formed even if the rate of chlorophyll(ide) formation does not exceed the rate in seedlings not treated with ALA.

These observations have already been made in principle by previous investigators (e. g. [239]). SISLER and KLEIN [239] discovered that incubation of dark-grown bean leaves on ALA (0.01 M) resulted in a 10-fold increase of the protochlorophyll peak (at 628 nm, in acetone). They further noticed that chlorophyll (that is, material absorbing at 665 nm, in acetone) synthesized in the presence of ALA is apparently destroyed by light. Fig. 183 indicates that the interpretation of this effect as given by SISLER and KLEIN is probably not correct. They suggested that the photodestruction of chlorophyll in ALA-treated tissue might be due to the failure of other components of the photosynthetic apparatus to develop

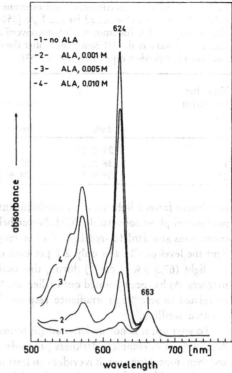

Fig. 183. Absorption spectra of acetone extracts from far-red-grown mustard seed-lings, incubated for one hour (47 to 48 h after sowing) in solutions of δ-amino-levu-linic acid (ALA). The technique of incubation was described previously [143]. Onset of far-red light: 36 h after sowing. Extraction: 60 h after sowing. For determination of the spectra the pigments were transferred into ether (after [129])

at the same rate as the chlorophyll, resulting in photodestruction of the chlorophyll formed. Rather, we must conclude that ALA exerts a destructive effect on the plastid with the consequence of artifacts, including lability of chlorophyll(ide) towards light. However, the artifacts caused by ALA incubation seem to be quite general. As an example, the phytochrome contents of mustard cotyledons as detected by the "Ratiospect" decreased after an incubation on 0.0025 M ALA by more than 50 per cent [128].

c) Control of Protochlorophyll (PChl) and Chlorophyll a (Chl a) Accumulation by Phytochrome

Does P_{fr} exert any control over re-synthesis and pool size of photoconvertible protochlo-rophyll after a saturating PChl → Chl a photoconversion has occurred? Table 21 clearly shows that $P_{fr \ (ground \ state)}$ does indeed control the rate of PChl synthesis as well as the size of the PChl pool. The results in the columns "red" and "red + far-red" confirm those obtained recently for maize tissue by Spruit and Raven [243]. The column on the right is of especial interest. While 5 min of the standard far-red light does not lead to a detectable decrease of the amount of photoconvertible PChl, synthesis and pool size of PChl are obviously affected by this treatment. This result can only be attributed to the P_{fr} which was produced by the far-red light. We recall that the photostationary state of the phytochrome system which is established by standard far-red light contains about 3 per cent P_{fr} (cf. 7th lecture) and we recall also that in some responses of the mustard seedling the degree of the response induced by 5 min far-red light is at least 50 per cent of the response induced by 5 min red light (cf. Fig. 11, Table 2). Fig. 184 confirms that

Table 21. Results of induction-reversion experiments with the mustard seedling. Standard red light [67.5 μW · cm⁻²] and standard far-red light [350 μW · cm⁻²] were given for 5 min each at 48 h after sowing (25°C). The amount of photoconvertible protochlorophyll (PChl) was determined according to DE GREEF et al. [83] immediately after the light treatment and 12 or 24 h later. The values are means of 4 independent experiments (after [153])

Time after Irradiation [h]	PChl [p moles/pair of cotyledons]			
	dark	red	red + far-red	far-red
0	28 ± 4	0	0	28 ± 4
12	38 ± 2	52 ± 6	36 ± 4	50 ± 4
24	38 ± 4	74 ± 10	42 ± 4	58 ± 6

continuous far-red light (that is, predominantly P_{fr}*) leads to a considerable increase in pool size of photoconvertible PChl. Nevertheless, the rate of accumulation of Chl a under continuous standard far-red light (●) is very small. At 36 h after the onset of far-red light the level of Chl a is only 2.2 per cent of the level reached under continuous weak red light (67.5 μW · cm⁻²) during this period. Chlorophyll b could be detected only in traces. As has been pointed out earlier (cf. 4th lecture) the very low level of Chl a cannot be related to any "high irradiance response" mediated by standard far-red light in the mustard seedling.

To sum up, it is not understood (in terms of developmental physiology) why plastid development in complete darkness proceeds only to a certain stage (proplastid, etioplast) and then stops. However, it is evident (at least in principle) that the light-dependent resumption of plastid development is mediated by the active phytochrome, P_{fr}. Phytochrome

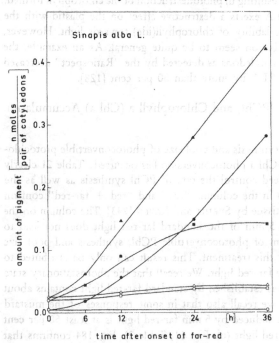

Fig. 184. Time-course of chlorophyll a and protochlorophyll accumulation in the cotyledons of the mustard seedling. The amount of photoconvertible protochlorophyll was determined according to DE GREEF et al. [83]. Onset of continuous far-red light: 36 h after sowing. The values are means of 4 independent experiments (after [153])

● , chlorophyll a, without white light treatment

▲ , chlorophyll a, after white light treatment (10 min)

■ , photoconvertible protochlorophyll(ide) (PChl) (▲ minus ●) open symbols, dark control; solid symbols, far-red treated

probably controls every course of biosynthesis of plastid components. As far as free energy and building blocks are concerned, the developing plastid calls upon the rest of the cell for many constituents. This heterotrophy includes not only nutritional material but very probably also complicated molecules such as enzymes. The second point where light comes into play is the phototransformation of protochlorophyllide to chlorophyllide a. A controlling function of P_{fr} is obvious even in this process but the "mechanism" is still obscure. There are no indications so far that the PChl $\rightarrow$ Chl a transformation acts as a signal for the resumption of biosynthesis of plastidal or extraplastidal components.

Suggested Further Reading

BOGORAD, L.: Biosynthesis and morphogenesis in plastids. In: Biochemistry of Chloroplasts, Vol. II (T.W. GOODWIN, edit.). London: Academic Press 1967.

GOODWIN, T.W.: Terpenoids and chloroplast development. In: Biochemistry of Chloroplasts, Vol. II (T.W. GOODWIN, edit.). London: Academic Press 1967.

GOODWIN, T.W.: Biosynthesis by chloroplasts. In: Structure and Function of Chloroplasts (M. GIBBS, edit.). Berlin-Heidelberg-New York: Springer 1971.

HÄCKER, M.: Phytochrome control of degradation of storage protein and formation of plastids in the cotyledons of the mustard seedling (*Sinapis alba* L.). Planta 76, 309 (1967).

SCHIFF, J.A.: Developmental interactions among cellular compartments in *Euglena* in control of organelle development. Soc. Exptl. Biol. Symp. 24, 277 (1970).

Appendix: Phytochrome-mediated Control of Peroxisome Enzymes

"Molecular" models to describe developmental genetics in higher organisms (e. g. [26]) use the assumption that several enzymes can be induced (or repressed) simultaneously and coordinately. The following example suggests that this assumption is justified. Two enzymes which are thought to be functionally related marker enzymes of the peroxisomes [271] were selected for investigation: glycollate oxidase (a flavoprotein which catalyzes the oxidation of glycollate to glyoxylate, using molecular oxygen, cf. Fig. 83) and glyoxylate reductase, an NAD-linked enzyme.

A comparison of the induction patterns of glycollate oxidase (Fig. 185) and glyoxylate reductase (Fig. 186) suggests that both enzymes are induced and regulated simultaneously and coordinately in the mustard seedling. While the factors which control the enzyme kinetics in complete darkness are unknown, the light effect (continuous standard far-red light) can be attributed to P_{fr} (cf. 4th lecture).

The question arose as to whether or not the enzyme kinetics in Fig. 185, 186 reflect the control by P_{fr} over compartment (that is, peroxisome) biogenesis. In this case one can assume that the actual amount of enzyme is determined by the availability of space (or sites) in the growing compartment. The alternative would be that P_{fr} controls enzyme synthesis as such, while the formation of the compartment is not significantly influenced by P_{fr}. In this case one has to assume that the newly-formed enzymes are transferred from the cytoplasm into preexisting particles. In an effort to decide between these alternatives the kinetics of the enzyme catalase were investigated. Catalase is one of the enzymes which characterize isolated peroxisomes (but unfortunately also glyoxysomes). If P_{fr} controls the biogenesis of peroxisomes (and thus affects enzyme accumulation) it is to be

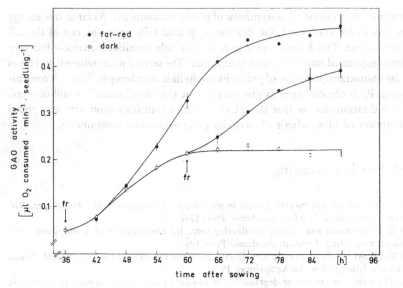

Fig. 185. Time-course of increase of glycollate oxidase (GAO) in the mustard seedling in the dark and under standard far-red light. The onset of far-red light is indicated by the arrows (after [205])

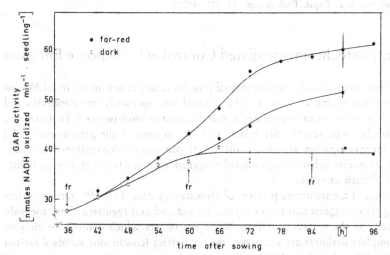

Fig. 186. Time-course of increase of glyoxylate reductase (GAR) in the mustard seedling in the dark and under standard far-red light. The onset of far-red light is indicated by the arrows (after [205])

expected that the control of catalase activity will be similar to the control found with glycollate oxidase and glyoxylate reductase. This is obviously not the case (cf. Fig. 88). The time-course of catalase activity in the mustard seedling is nearly the same in darkness and under continuous far-red light.

However, it is possible that only a small fraction of the total catalase is related to the peroxisomes while the larger part is localized in the glyoxysomes. In this case the slight difference between the dark and far-red kinetics in the second part of the time-course (cf. Fig. 88) could be meaningful. Since it has only recently become possible to separate the catalase isoenzymes of the mustard seedling by starch gel electrophoresis, the problem is still open for investigation.

The information about the regulation of microbody enzymes (peroxisomes and glyoxysomes) might even be relevant for plastid biogenesis. In the mustard cotyledons the microbodies, as localized by staining the site of catalase activity, originate in close proximity to the plastids, and sometimes the microbodies appear to be appressed to the side of the plastids [54]. A comparison of Fig. 185, 186 and 179 further indicates that the formation of *peroxisomes* in the mustard cotyledons is closely correlated with the formation of plastids.

Besides malate synthetase, isocitrate-lyase is a typical marker enzyme of the glyoxysomes. Fig. 87 shows that P_{fr} (operationally, continuous far-red light) has no influence whatsoever on the time-course of isocitrate-lyase activity in the cotyledons of the mustard seedling. This fact very probably means that P_{fr} does not control the time-course of those enzymes which are constituents of glyoxysomes. Since the time-course of total catalase activity (cf. Fig. 88) is similar to the time-course of isocitrate-lyase, it can be assumed that most of the catalase measured by our assay is a glyoxysomal constituent while only a small part is located in the peroxisomes.

In conclusion, the data which are available at present suggest that the formation of peroxisomes is under the control of phytochrome (in some way or other) while the formation and decay (?) of glyoxysomes is not. However, the experimental information available at present from the mustard seedling is restricted to time-courses of marker enzymes and is thus limited. A correlative biochemical and ultrastructural study of the time-course of microbody formation and disappearance (?) has not yet been attempted in our system. Such a study has recently been undertaken in cucumber cotyledons [275]. Unfortunately, only high irradiance *white* light (8000 lx) was used as an experimental variable. However, the principal results and conclusions can possibly be extrapolated to the mustard seedling. It was found that although glyoxysomal enzyme activities drop rapidly while peroxisomal enzyme activities increase rapidly during the transition period in the white light, the electron microscope evidence does not indicate that glyoxysomes are being degraded or peroxisomes being formed. It was concluded that the developmental transition from glyoxysomal to peroxisomal function almost certainly does not involve the actual replacement of one population of microbodies by another. Rather, the evidence suggests that at least some of the glyoxysomes may acquire peroxisomal function during the course of development through a change in enzyme complement.

Suggested Further Reading

DRUMM, H., FALK, H., MÖLLER, J., MOHR, H.: The development of catalase in the mustard seedling. Cytobiologie 2, 335 (1970).

KAROW, H., MOHR, H.: Changes of activity of isocitritase (EC 4.1.3.1.) during photomorphogenesis in mustard seedlings. Planta 72, 170 (1967).

TOLBERT, N.E.: Microbodies – peroxisomes and glyoxysomes. Ann. Rev. Plant Physiol. 22, 45 (1971).

TRELEASE, R.N., BECKER, W.M., GRUBER, P.J., NEWCOMB, E.H.: Microbodies (glyoxysomes and peroxisomes) in cucumber cotyledons. Plant Physiol. 48, 461 (1971).

Phytochrome and Flower Initiation

1. Definitions

In many plants the formation of flowers depends on the length of the daily light period (Photoperiodism). The light stimulus is perceived by the leaves. The phenomenology of photoperiodism is somewhat complicated, since many deviations from the general rules have been described (e.g. [280]). However, since we are restricting our interest to the principles of photoperiodism, we can restrict our considerations to obligatory short-day plants (SDP) and obligatory long-day plants (LDP). An SDP will only flower if it has received a minimum number of short-days (e.g. 8 h light per day). If the plant is kept continuously under long-day conditions (e.g. 16 h light per day), no flowering occurs. An LDP will only flower if it has received a minimum number of long-days (e.g. 16 h light per day or continuous light). If the LDP is kept exclusively under short-day conditions the plant will not flower.

The term critical daylength has sometimes been used to define SDP and LDP: SDP initiate flowers if the daily photoperiod is shorter than a certain critical daylength, and LDP initiate flowers if the daily photoperiod is longer than a certain critical length. For example [280], both chrysanthemum (SDP) and henbane (LDP) will initiate flowers in a 12 to 14 h daily photoperiod. The difference is that the SDP chrysanthemum does not flower if the daily photoperiod (=daylength) is longer than the critical 14.5 h.

It was shown in SDP as well as in many LDP that the effect of a long-day treatment can be replaced by the following program: short-day main light period (e.g. 8 h white fluorescent light of high irradiance per day) plus a relatively brief irradiation (e.g. 5 min) around the middle of the corresponding dark period (night interruption) (Fig. 187). This means that SDP require long nights for flower initiation whereas in LDP long nights are inhibitory to flower initiation.

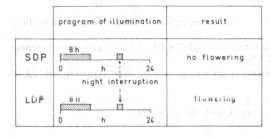

SDP = short day plant
LDP = long day plant

Fig. 187. This scheme is to illustrate the fact that in many cases the program "short-day main light period + night interruption" can replace a long-day treatment with respect to flower initiation. This is true for short-day plants as well as for many long-day plants

2. Involvement of Phytochrome

Red light falling on the leaves was early found to be highly effective in preventing flowering of plants requiring long nights for flower initiation. Action spectra for suppression of flowering by night interruptions were measured by the Beltsville group and found to be almost the same for two short-day plants, *Xanthium pennsylvanicum* (cocklebur) and *Glycine max* (cv. Biloxi, soybean) [196]. The action spectra were characterized by a strong peak of action in the red part of the spectrum and only a small effect in the blue. Action spectra for initiation of flowering in long-day plants (*Hordeum vulgare*, cv. Wintex, barley; *Hyoscyamus niger*, henbane) were also measured by interrupting long nights with light of different wavelengths. The results showed that the action spectra for LDP are almost the same as the ones for SDP. It had to be concluded that the same photoreceptor pigment is involved in the two types of plants even though the flowering response is different – suppression in SDP and initiation in LDP [104]. The course of the action spectrum and the fact that the action of red light can be fully reversed by following the red light immediately with a corresponding dose of far-red light clearly show that phytochrome is involved in the photoperiodic response (cf. Fig. 13).

3. Interaction between Phytochrome and Endogenous Rhythms

The effects of brief light exposures on flowering in certain plants depend on the time at which they follow a transition from light to darkness (main light period → dark transition). At the end of the main light period (e.g. 8 h of high irradiance white fluorescent light), phytochrome is mainly (approximately 80%) in the form P_{fr}. The question was whether or not an SDP would respond to a brief irradiation with far-red light at the end of the main light period. This is indeed the case. The work done by NAKAYAMA [185] with *Pharbitis nil* has often been quoted in this context. Seedlings of *Pharbitis nil* can be induced to flower as soon as the cotyledons unfold. *Pharbitis nil* plants growing in 2 to 8 h light periods in white light of high irradiance and 22 to 16 h dark periods, can be *potentiated* to flower if P_{fr} is present at maximum concentration at transition to darkness. Therefore a brief irradiation with far-red light at the end of the main light period results in an *inhibition* of flowering. This effect can be reversed by red light. The action of the same red light on *Pharbitis nil* 6 to 10 h after the beginning of the dark period is opposite to that at the beginning: flowering is suppressed [186].

The experimental information available allows the general conclusion that in SDP the presence of P_{fr} is necessary at the beginning of the long night period in order to make flower initiation possible. However, if P_{fr} is present in the middle of the night period, it inhibits flower initiation [41, 63]. Since the reverse phenomenon has been observed in LDP (P_{fr} inhibits flowering at the beginning of the night period but stimulates flowering if applied later [279, 280]) the conclusion that one is dealing with a change of sensitivity of the system (leaves) towards P_{fr} in both types of plants was eventually inevitable. However, the idea that the extent and even the quality of the effect of P_{fr} on flowering is determined by an endogenous rhythm has been accepted only reluctantly by the scientific community [29, 103, 62, 110, 43].

The term endogenous rhythm is used to designate any rhythmic changes of activity in a living system which persist even under constant conditions. Since the length of a period under constant conditions may deviate somewhat from 24 h, the rhythm is called circadian. The agent behind the endogenous rhythms has been called the "Physiological Clock" by BÜNNING [29]. The molecular nature of the physiological clock is still unknown.

The photoperiodic reaction is probably determined by an interaction between an endogenous circadian rhythm and phytochrome. Both components are essential for the photoperiodic response. The endogenous rhythm determines the changing sensitivity of the plant to P_{fr}, involving both quantitative and qualitative changes (photophil and photophobe phases with respect to P_{fr} and with respect to flowering). Using the short-day plant *Chenopodium amaranticolor*, KÖNITZ [138] has shown in a series of impressive experiments that during the photophil phase of the plant, as indicated by leaf movements, the *absence* of P_{fr} is inhibitory to flowering, whereas during the photophobe phase the *presence* of P_{fr} is inhibitory to flowering. The highest sensitivity to red light (i.e. to P_{fr}) is not in the middle of the dark period but in the middle of the photophobe phase as indicated by the diurnal leaf movements. Correspondingly the highest sensitivity to far-red light (i.e. the highest requirement for P_{fr}) is to be found not in the middle of the main light period, but in the middle of the photophil phase. The characteristic change of sensitivity towards light during the dark period could also be demonstrated in experiments with longer dark periods (up to 72 h) (Fig. 188, 189).

To sum up, the transformation of P_r to P_{fr} is used by the plant to determine if there is light or darkness at any point of the endogenous diurnal rhythm. If white light is falling on the plant, P_{fr} is formed. The status of the endogenous rhythm then determines whether this P_{fr} promotes or inhibits flowering [160].

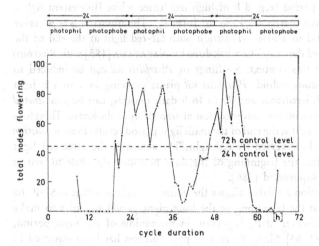

Fig. 188. Flowering response of Biloxi soybean to night interruptions given at different times during the dark period of a 72 h cycle. The control plants were exposed to seven cycles, each cycle consisting of an 8 h photoperiod followed by a 64 h dark period. In the various treatments the 64 h dark period was interrupted at various times by illumination of 1 h duration. Flowering of the controls with no night interruptions for 72 h cycles as well as for 24 h cycles is represented by horizontal lines. Each point on the curve represents a treatment with the interruption beginning at that time. Alternating photophil and photophobe phases are shown on the top of the figure. The participation of an endogenous rhythm in this photoperiodic response, with alternate 12 h photophil and photophobe phases, is clearly indicated (after [91])

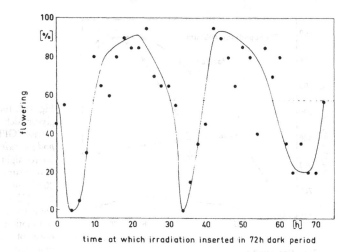

Fig. 189. The effect of a night interruption with 4 min red light given at various times (2 h intervals) during a 72 h dark period on the flowering of *Chenopodium rubrum* (SDP). Plants were kept in continuous high irradiance white light before and after the single 72 h dark period. Horizontal line is 72 h dark control (57% flowering) (modified after [42])

De novo synthesis of P_r as well as the quick disappearance of P_{fr} in darkness (through decay or through reversal to P_r; cf. Fig. 22) are essential for the controlling function of phytochrome in photoperiodism. These properties of the phytochrome system enable the plant to use it repeatedly to detect whether or not there is light falling on the plant. The rate of disappearance of P_{fr} controls how precisely a plant can determine the end of a light period. If P_{fr} were stable, the phytochrome system could not be used by the plant in the photoperiodic response.

The measurement of time has been found to be very precise in some plants. An example: With repeated cycles, the response in the SDP *Xanthium pennsylvanicum* changes from zero flowering to 100 per cent flowering as the daily dark period is lengthened by 30 min [227].

Endogenous rhythms determine the sensitivity of a plant (or of systems within the plant) towards P_{fr}. This is true for flowering and also for any other response. In order to detect the molecular basis of the interaction between P_{fr} and the physiological clock (rather than its multiple manifestations, the rhythms), several investigators have been using simpler responses than flowering. Some of these data will now be presented as a typical example of this new approach, which hopefully will lead to an understanding of the manner in which the physiological clock and P_{fr} interact during the development of a living system.

E. WAGNER[1] and S. FROSCH[1], working with seedlings of *Chenopodium rubrum*, have recently been able to study the induction of the NADP-linked glyceraldehyde-3-phosphate dehydrogenase (GPD) by continuous far-red light with and without the interaction of an endogenous rhythm [284]. In one experiment the seeds were induced to germinate in the dark by an acid treatment and were kept continuously in the dark at precisely 20°C. After 3 days the dark-grown seedlings were placed under standard far-red light and the kinetics of enzyme induction were measured (Fig. 190). It was found that formation of

_____

1 Permission to quote these data in the present context is gratefully acknowledged.

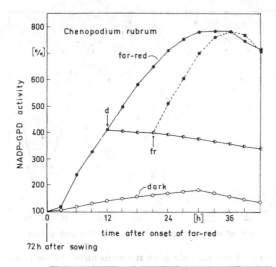

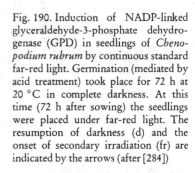

Fig. 190. Induction of NADP-linked glyceraldehyde-3-phosphate dehydrogenase (GPD) in seedlings of *Chenopodium rubrum* by continuous standard far-red light. Germination (mediated by acid treatment) took place for 72 h at 20 °C in complete darkness. At this time (72 h after sowing) the seedlings were placed under far-red light. The resumption of darkness (d) and the onset of secondary irradiation (fr) are indicated by the arrows (after [284])

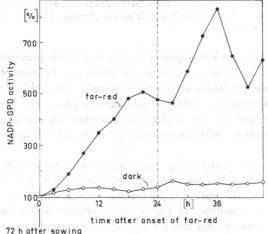

Fig. 191. Induction of NADP-linked glyceraldehyde-3-phosphate dehydrogenase (GPD) in seedlings of *Chenopodium rubrum* by continuous standard far-red light. Germination took place for 72 h in complete darkness at alternating temperatures (32.5° : 10 °C) (after [284])

GPD in the dark is very modest; however, a strong enzyme synthesis can be evoked by continuous far-red light. Enzyme increase requires the continuous presence of P_{fr} (operationally, standard far-red light), since enzyme increase stops instantaneously if the far-red light is turned off. A secondary irradiation with the same far-red light leads to an immediate resumption of apparent enzyme synthesis. No rhythmic phenomena are detectable; thus far the situation is very similar to what we observe in the mustard seedling under precisely-controlled, constant conditions of temperature and humidity (or water potential) during sowing, seed germination and seedlings' growth. However, when dark germination is evoked by cyclic temperature changes (32.5° versus 10°C) instead of treatment with acid at constant temperature, the kinetics of the enzyme activity oscillate in the dark as well as under continuous standard far-red light (Fig. 191). The interpretation is that the treatment with temperature cycles during seed germination has released the endogenous rhythm, while the induction of the NADP-GPD by far-red light after germination at constant temperature takes place in a system without endogenous rhythms.

It is obvious that measurements of enzyme induction in systems which show endogenous rhythms are only meaningful (in a quantitative sense at least) if the experimental conditions are so precisely controlled that the endogenous rhythm does not interfere with the system which produces or decays the enzyme in question. This is the case, for example, with the mustard seedling under the experimental conditions used so far during sowing of the seed, seed germination and seedlings' growth. However, if a mustard plant is kept under rhythmic conditions, e.g. under short-day conditions, an endogenous rhythm shows up which can be demonstrated impressively in connection with flower initiation by a night interruption (Fig. 192). The mustard plant is an LDP, and flower initiation can be evoked in accordance with the general scheme presented previously (cf. Fig. 187).

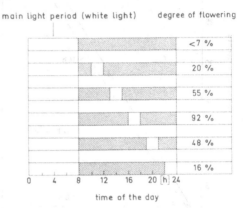

Fig. 192. The effect of a night interruption with 2 h blue light (cf. Fig. 197) given at various times during a 16 h dark period on the flowering of Sinapis alba (mustard, LDP). Treatment was repeated for 6 consecutive days. Plants were kept under short-day conditions (8 h main light period with standard white light) before and after the 6 day period of treatment with night interruptions (after [92])

4. The "Primary Action of P_{fr}" in Connection with Flower Initiation

We have already mentioned that the SDP *Pharbitis nil* can be induced to flower by a short-day treatment as soon as the cotyledons unfold [185]. However, if a brief irradiation with red light is given close to the middle of the dark period, flower initiation is suppressed. This red light effect is only reversed by far-red light if the red light is applied within a few seconds, followed within half a minute by intense far-red irradiation for a few seconds [105, 70]. This fact indicates that P_{fr} action in this particular response must have taken place in less than a minute. Formally, this has been interpreted by the assumption that the primary reactant for P_{fr} (X) is present in great excess [105].

5. Photoperiodic Effects on Vegetative Characteristics

Although less well documented than flower initiation, other responses to photoperiods appear to be controlled by phytochrome in a similar manner. The prevention of dormancy in buds by sustained long-day treatment is of special interest (Fig. 193). In most perennial species the buds do not go into dormancy if long-day conditions are maintained [188]. While dormancy (including the cessation of internodal growth) is a response occurring at the apex with the formation of resting buds or the shedding of the terminal bud, the light stimulus (the length of the daily photoperiod) is perceived by the leaves. Thus in flower initiation as well as in the vegetative photoperiodic responses, chemical messengers appear to be formed in the leaves in particular photoperiods and are translocated about the plant to the sites of the morphological responses. Unfortunately the chemical nature of these messengers is unknown. This is true for the hypothetical "flowering hormone" (florigen) as well as for the messengers of the vegetative photoperiodic responses.

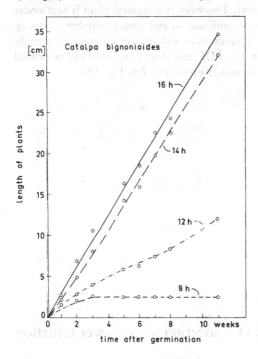

Fig. 193. Vegetative growth of young plants of *Catalpa bignonioides* at four different photoperiods (16, 14, 12, or 8 h light per day) (after [50])

Suggested Further Reading

BÜNNING, E.: The Physiological Clock. New York: Academic Press 1963.
CUMMING, B.G., WAGNER, E.: Rhythmic processes in plants. Ann. Rev. Plant Physiol. *19*, 381 (1968).
EVANS, L.T. (edit.): Environmental Control of Plant Growth. New York: Academic Press 1963.
EVANS L.T. (edit.): The Induction of Flowering. Ithaca, N.Y.: Cornell University Press 1969.
HAMNER, K.: Endogenous rhythms in controlled environments. In: Environmental Control of Plant Growth (L.T. EVANS, edit.) New York: Academic Press 1963.
HENDRICKS, S.B., BORTHWICK, H.A.: The physiological functions of phytochrome. In: Chemistry and Biochemistry of Plant Pigments (T.W. GOODWIN, edit.). London: Academic Press 1965.
HILLMAN, W.S.: Photoperiodism and vernalization. In: Physiology of Plant Growth and Development (M.B. WILKINS, edit.). London: McGraw-Hill 1969.
VINCE, D.: The control of flowering and other responses to daylength: part I, II. J. Royal Horticult. Soc. *45*, 214 (1970).

Phytochrome and Seed Germination

1. Operational Criteria for the Involvement of $P_{fr \ (ground \ state)}$ in Seed Germination [21]

If lettuce seeds of a light-requiring variety (*Lactuca sativa*, cv. Grand Rapids) are sown on a suitable medium at 25° C and placed in darkness, most of the seeds do not germinate. One minute of red light induces 100% germination. If the red light is followed with far-red light, the induction of germination is nullified (cf. Fig. 6). The explanation is that the red treatment establishes a photostationary state where about 80 per cent of the total phytochrome is present as P_{fr}. Under these conditions all seeds start to germinate. The far-red treatment, on the other hand, leads to a photostationary state where only a few per cent of the total phytochrome is present as P_{fr}. Under these conditions germination cannot proceed since the P_{fr} requirement (threshold level of P_{fr}) is not satisfied. The effect of $P_{fr \ (ground \ state)}$ within an individual seed must be understood in terms of a threshold (or all-or-none) response. The distribution of light-sensitivity of a seed population (per cent germinating seeds as a function of the light dose applied) can be understood as a distribution of the threshold values for $P_{fr \ (ground \ state)}$ within the seed population. Applied to Fig. 6, this means that after 1 min of red light, the threshold level of $P_{fr \ (ground \ state)}$ is exceeded in all the seeds; after 1 min of far-red light, on the other hand, the threshold level is exceeded only in a small fraction of the seed population. It is assumed that those seeds which germinate in complete darkness contain or form some P_{fr} even without light. This conclusion is suggested by the fact that the percentage of dark germination can sometimes be decreased slightly by a brief treatment with far-red light (cf. Table 22).

2. The Inhibitory Effect of Long-term Far-red Light on Seed Germination [169]

It was mentioned in the previous section that in some light-requiring lettuce seed full germination can be induced by a brief irradiation of the imbibed seeds with red light, and that this induction can be fully reversed by immediately following with a corresponding dose of far-red light. However, if a considerable period of time has elapsed before the far-red light is applied (e.g. 12 h), no reversion will occur. The interpretation of this fact is that the process of germination has escaped from the control by P_{fr} [21]. This, however, is only part of the story. If high doses of far-red light (e.g. 256 min instead of 1 or 4 min) are applied, germination can still be completely suppressed even in a lettuce seed population (cv. Grand Rapids, tip burn resistant strain) which has nearly escaped from the control by a short-term irradiation with far-red light. 1 min of red light given after the prolonged far-red treatment restores the potential for full germination; 1 min of far-red light after this red pulse suppresses germination completely, etc. (Table 22). A further important feature is that even dark germination is suppressed by long-term irradiation with far-red light. Using the technique of simultaneous irradiation with two wavelengths (cf. Fig. 56), HARTMANN [94] demonstrated that the inhibitory effect of prolonged far-red light on lettuce seed germination (cv. Grand Rapids) must be attributed to phytochrome as well. A hypothesis for explaining these facts requires the assumption that P_{fr} can originate in these seeds even in complete darkness. BOISARD et al. [18, 19] have shown, working with a non-light-requiring variety of lettuce seed (cv. Reine de Mai), that P_{fr} is indeed formed from P_r in complete darkness ("inverse reversion"). The inhibitory effect of a prolonged treatment with far-red light on seed germination in the variety Reine de Mai can be explained by the assumption that the far-red light maintains a P_{fr} level *below* the threshold level which must be exceeded in the seeds over a long period in order to keep the process of germination going. In complete darkness the threshold level of P_{fr} can be exceeded by "inverse reversion" ($P_r \rightarrow P_{fr}$). The responses of the lettuce seeds of the variety Grand Rapids (cf. Table 22) can be understood as follows (Fig. 194). The germination process is set in motion by a threshold level of P_{fr} which is relatively high and which can only be reached in the majority of the seeds of a particular seed population with the help of red light (short-term irradiation, cf. Table 22). Once the germination process has started

Table 22. The inhibition of lettuce seed germination (dark germination and germination potentiated by $P_{fr \, (ground \, state)}$) by a prolonged treatment with far-red (fr) light (after [169])

Program of Irradiation	Per cent Germination
dark	36
1 min fr	31
1 min red	97
1 min red – 1 min fr	29
1 min red – 4 h dark – 4 min fr	85
1 min red – 4 h dark – 64 min fr	52
1 min red – 4 h dark – 64 min fr – 1 min red	97
1 min red – 4 h dark – 64 min fr – 1 min red – 1 min fr	54
1 min red – 4 h dark – 256 min fr	6
1 min red – 4 h dark – 256 min fr – 1 min red	95
1 min red – 4 h dark – 256 min fr – 1 min red – 1 min fr	3

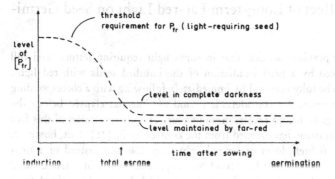

Fig. 194. A scheme to illustrate the hypothesis which explains the germination behavior of a light-requiring seed of lettuce (cv. Grand Rapids). The requirement for P_{fr} decreases quantitatively as the pre-germination processes take place in the seed. However, there is a continuous requirement for a low level of P_{fr} until the breakthrough of the radicle

(escape), the continuous requirement for P_{fr} is similar to the one in the variety Reine de Mai. This requirement can be satisfied in all seeds by inverse reversion, that is, dark formation of P_{fr}. A related phenomenon is that in many seeds (e.g. lettuce seed, cv. Great Lakes, or tomato seed) which readily germinate in darkness, a requirement for red light can be "induced" by holding them for 24 h or more under continuous far-red light. This is probably the same effect which was noticed in Table 22 with respect to the dark-germinating part of the seed population. The interpretation favored by the Beltsville group [104] is that low levels of P_{fr} maintained by continuous far-red irradiation drain off the substrate for P_{fr} action without establishing an adequate product level for germination. However, it seems more plausible to explain the "induced red light requirement" by the assumption that inverse reversion has ceased during the treatment with far-red light.

The effect of light on germination of those seeds which germinate readily in the dark but are inhibited by prolonged illumination even by white light (e.g. *Nemophila insignis* [220], or *Amaranthus caudatus* [134]) can possibly be understood as follows: P_{fr} is required at a relatively high level for germination to proceed. This P_{fr} threshold level can be exceeded in complete darkness *via* inverse reversion. White light ($[P_{fr}]/[P_{tot}] > 0.5$) is inhibitory because the decay rate of P_{fr} is too high under these conditions (as compared to the rate of *de novo* synthesis of phytochrome), and thus the seeds become rapidly depleted of phytochrome. Eventually the seeds go into dormancy (light-mediated secondary dormancy). The very strong inhibitory effect of far-red light on germination of this type of seed [218, 69] can be explained as in the case of lettuce or tomato seed: The level of P_{fr} which is maintained by far-red light, is too low for germination to proceed. However, the photostationary amount of P_{fr} is high enough to drain off the substrate for P_{fr} action even though the threshold for germination is not reached. This type of explanation implies the assumption of a decay of seed phytochrome which is possibly not justified (cf. next section). However, until and unless the stability of phytochrome is demonstrated experimentally in those seeds whose germination is inhibited by white light, the explanation given seems to be the most plausible one. According to TAYLORSON and HENDRICKS [269] the germination behavior of the seeds of *Amaranthus retroflexus* can only be understood if net synthesis of P_r in the dormant seed is assumed as well as P_{fr} destruction. It was concluded that synthesis of P_r can proceed in the seed even though germination is blocked by lack of P_{fr}.

The question of whether or not a "high irradiance response" exists in light-controlled seed germination has been neglected so far in the present discussion. This omission is probably not justified. A number of data on seed germination indicate that a "high irradiance reaction" exists which stops germination even at a very late stage in the process, possibly by direct inhibition of radicle growth. Dual wavelength experiments which followed HARTMANN's approach [94] clearly indicate that the "high irradiance" photoinhibition of germination in *Amaranthus caudatus* seed can also be understood in terms of phytochrome [69]. However, the manner in which P_{fr}^* is involved in control of the germination process requires further investigation.

3. Direct Phytochrome Measurements in Seed

It was concluded in the previous section that at least in some seed P_{fr} can be formed in complete darkness. An "inverse reversion" ($P_r \rightarrow P_{fr}$) has been demonstrated in seed by spectrophotometric measurements [19, 135, 242]. The P_{fr} found in a dark-grown seed could have originated *via* inverse reversion from P_r during the imbibition of the seed, or one is dealing with P_{fr} which was maintained by light during the maturation of the seed and preserved as P_{fr} during the rapid dehydration of the ripening seed. This latter suggestion must be taken seriously since WALTER and SHROPSHIRE [287] have shown with *Arabidopsis thaliana* that the ability of the seed to germinate in complete darkness depends on the ratio (red light/far-red light) under which the *mother plant* was kept. If the ratio was high, the seeds will germinate in complete darkness; if the ratio was low, however, the seeds will require light for germination. Some results obtained with *Amaranthus caudatus* indicate [135] that even after a long-term irradiation with red light a significant destruction of the original seed phytochrome does not seem to occur ("stability of seed phytochrome"). Moreover, the spectral properties of the seed phytochrome do not coincide completely with those of the newly-formed seedling phytochrome [242]. However, this observation might be due to hydration effects in the tissue [219]. In any case, if there are properties peculiar to the seed phytochrome ("inverse reversion"; "stability"), they can no longer be detected in the seedling. The reason might be that the seedling's phytochrome which is synthesized *de novo* (starting during the course of the germination process) does not have these properties and will eventually outnumber the original seed phytochrome by many orders of magnitude (cf. Fig. 27). Another alternative should not be overlooked: In seeds of white-seeded mustard (*Sinapis alba*) no P_{fr} could be detected in complete darkness [148]. Furthermore, the germination process of the mustard seed is not influenced by light; even far-red light at high irradiances is without effect. It is very probable that the germination process in mustard is not under the control of P_{fr} at all, whereas the development of the seedling is (cf. Fig. 1). The appearance of phytochrome sensitivity in the seedling's morphogenesis depends on the temporal development of the pattern of primary differentiation which determines the pattern of competence, that is, the response pattern of the seedling towards P_{fr} (cf. 14th lecture).

4. The "Mechanism" of Germination

Phytochrome was first detected physiologically in studies on light-mediated seed germination. However, the studies on the molecular mechanism of phytochrome action in seed germination have proceeded very slowly, and even at present there is no agreement about the action mechanism of phytochrome in seed germination.

The events during seed germination are comparable to the events which occur during the early development of many animal embryos, e. g. the sea urchin embryo. This is amazing insofar as the seeds contain multicellular, differentiated systems, namely young sporophytes, whose dormant state was only transient. Two examples: In the wheat grain protein synthesis will start at the time of imbibition. However, during the first 24 h of the germination process, protein synthesis is exclusively programmed by such RNA as was already present in the dry seed [37].

In investigations with lettuce seed it was found that the process of germination cannot be inhibited by either Actinomycin D or Puromycin. Even if unreasonably high concentrations of these inhibitors are used, the radicle will only stop growing after a length of about 3 mm has been reached [255].

Experiments of this kind point to the conclusion that all the types of RNA and apparently all the enzymes which are required for the germination process are already present in the dormant seed.

To summarize: The question of how P_{fr} can make a lettuce seed germinate is still under debate. The attractive earlier concept that P_{fr} may set in motion a mechanism which causes the formation or activation of hydrolytic enzymes which in turn weaken the seed coat to such an extent that the radicle can break through, can obviously not be sustained.

SCHEIBE and LANG [251] concluded that light-requiring lettuce seeds do not germinate in the dark because the embryo cannot generate enough "growth potential" to overcome the mechanical restraining force of the external seed layers, in particular the endosperm. Red light (that is, P_{fr}) induces a "potential for growth" which enables the embryo to penetrate these layers. In a recent study, NABORS and LANG [182, 183] developed this concept further in terms of water potential. The experimental evidence indicates that P_{fr} causes a decrease in water potential in light-requiring lettuce seed which is equal to that which is required for germination. Thus, red-light-induced germination is the result of a decreased water potential in embryos. However, the "mechanism" of P_{fr} action, that is, the sequence of elementary steps starting with $P_{fr}X$ and eventually leading to the decreased water potential, is essentially unknown.

Suggested Further Reading

BOISARD, J.: La photosensibilité des akènes de Laitue (variété "Reine de Mai") et son interprétation par spectrophotométrie *in vivo* du photorécepteur. Physiol. vég. *7*, 119 (1969).

BORTHWICK, H.A., HENDRICKS, S.B., TOOLE, E.H., TOOLE, V.K.: Action of light on lettuce-seed germination. Bot. Gaz. *115*, 205 (1954).

KENDRICK, R.E., FRANKLAND, B.: Photocontrol of germination in *Amaranthus caudatus*. Planta *85*, 326 (1969).

KENDRICK, R.E., SPRUIT, C.J.P., FRANKLAND, B.: Phytochrome in seeds of *Amaranthus caudatus*. Planta *88*, 293, (1969).

ROLLIN, P.: Phytochrome, photomorphogénèse et photopériodisme. Paris: Masson 1970.

Examples of Blue-light-mediated Photomorphogenesis

1. Blue Light and the Phytochrome System

Both forms of phytochrome, P_r and P_{fr}, absorb in the blue and ultraviolet range of the spectrum (cf. Fig. 9). The extracted pigment has absorption maxima in this spectral region, P_r at about 370 nm and P_{fr} at about 400 nm. Therefore blue light must be expected to evoke responses similar to those caused by red and far-red light. While red light is 100 times more efficient in converting P_r and far-red light is 25 times more efficient in converting P_{fr} than blue light [206], photoequilibria can nevertheless be established. The ratio P_{fr}/P_{tot} as measured in vivo by HARTMANN and SPRUIT (cf. Fig. 10) is situated between about 20 and 40 per cent, depending on the particular wavelength in the range between 500 and 400 nm. Thus, P_{fr} will inevitably be formed if blue light is applied to a plant. There are, however, a considerable number of blue-light-dependent photoresponses where an explanation on the basis of phytochrome seems to be excluded. Our description will be restricted to four clear-cut examples.

2. Light-dependent Carotenoid Synthesis in Fusarium aquaeductuum

Carotenoid synthesis in this system can be regarded as a prototype of a photoresponse – including morphogenic responses – in fungi. The action spectrum (Fig. 195) shows that carotenoid synthesis is mediated only by light below 520 nm. The action spectrum has maxima at 375-380 nm and 450-455 nm, one shoulder at 430-440 nm and a further shoulder (or possibly a third maximum) between 470 and 480 nm. Carotenoids can be ruled out as possible photoreceptors, from this action spectrum. Rather, the spectrum resembles the absorption spectra of certain flavoproteins. Consequently the acting photoreceptor is thought to be a flavoprotein [214]. The causalities between the light absorption and the carotenoid accumulation are not fully understood so far. A good argument can be made, however, for believing that the light absorption causes the synthesis of enzymes which are required for carotenoid synthesis [214, 14].

Action spectra very similar to the one in Fig. 195 were determined for a multiplicity of responses in fungi. An example is light-mediated coremia-zonation in Penicillium claviforme [228]. Since no carotenoids could be detected in the mycelium of this fungus, and since diphenylamine (a potent inhibitor of carotenoid synthesis) did not exert any influence on the sensitivity of the mycelium towards light, it was concluded that the photoreceptor involved is a flavoprotein.

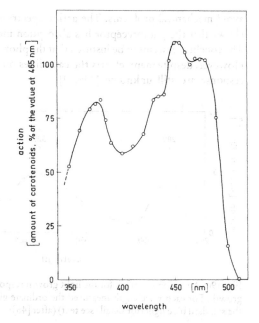

Fig. 195. An action spectrum of light-dependent carotenoid synthesis in the mycelium of the fungus *Fusarium aquaeductuum*. The amounts of carotenoids which can be induced by 4.2×10^{-7} Einsteins $\cdot$ cm^{-2} is given as a function of wavelength (after [214])

3. Polarotropism in Filamentous Germlings of the Liverwort *Sphaerocarpus donnellii*

This phenomenon has already been mentioned in the 3rd lecture. We recall (cf. Fig. 33) that the germlings will grow at an angle of 90° to the plane of vibration of the electrical vector of linearly polarized light which is applied from above. If the plane of vibration is turned, the sporelings will rapidly and correspondingly change their direction of growth. The action spectrum of the polarotropic response (cf. Fig. 34) shows that only short-wavelength light is effective. The details of the spectrum strongly suggest a flavoprotein as a photoreceptor. The photoreceptor molecules must be highly oriented in a dichroic structure, presumably close to the surface of the cell.

4. The Light Growth Response in the Sporangiophore of the Fungus *Phycomyces*

When sporangiophores are irradiated symmetrically about their long axis and at constant intensity, they grow at a uniform rate. This rate is the same regardless of the absolute intensity. If, however, the intensity is changed the rate of elongation changes transiently and finally returns to the initial value. This phenomenon has been called "light growth response". DELBRÜCK and SHROPSHIRE [46] determined an action spectrum of the light growth response using a null measuring system in which the spectral sensitivity was obtained by comparing the sensitivity at any wavelength to the sensitivity for a standard blue source. The symmetry of the irradiation was assured by rotating the specimen continuously at 2 rpm. This speed is sufficiently fast to eliminate tropic effects and sufficiently slow to avoid mechanical problems. The action spectrum of the light growth response (Fig. 196) shows that the photoreceptor has absorption maxima around 280, 385, 455 and 485 nm. The conclusion seems to be justified that the photoreceptor involved might be a flavoprotein. However, despite many efforts the causalities on the "molecular" level of the light growth response are still unknown [236, 9].

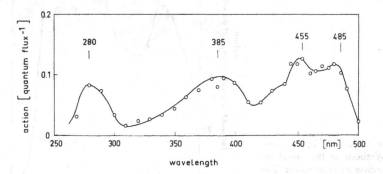

Fig. 196. An action spectrum for the light growth response in *Phycomyces* sporangiophore longitudinal growth. For each wavelength measured the ordinate gives the reciprocal quantum flux which matches the standard blue light (for details see text) (after [46])

5. General Conclusions

The action spectra mentioned (Fig. 34, 195, 196) point to a flavoprotein as the acting photoreceptor. Other reliable action spectra of blue-light-dependent photoresponses, e.g. those of phototropic responses in fungi and in higher plants, can be interpreted in the same way (cf. 22nd lecture). Therefore it seems plausible to suggest that the same photoreceptor is acting in all cases in which development and movement are mediated by light of short

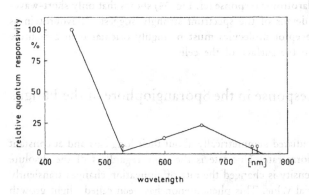

Fig. 197. A rough action spectrum of flower induction in mustard *(Sinapis alba)*, a long-day plant. Plants were kept for 28 days after sowing under short-day conditions (completely controlled environment (18 °C, 80% relative humidity); 8 h standard white light (8000 lx) per day = main light period). For another 6 days the plants received – besides the main light period – 2 h of supplementary monochromatic light in the middle of each dark period. After a further 14 days under standard short-day conditions the state of flower initiation was measured (after [92])

wavelengths. If indeed the physiologically effective short-wavelength light is absorbed in all systems by the same photoreceptor, the situation would be somewhat analogous to the one which has been discussed in connection with phytochrome: one and the same photoreceptor, diversity of photoresponses.

In those responses where phytochrome is involved it is still difficult to evaluate the contribution, if any, of a blue-light-specific photoreceptor. It is striking, however, that the short-wavelength part of the action spectrum for hypocotyl lengthening in lettuce (cf. Fig. 55) is very similar to the action spectra of responses where exclusively blue and ultraviolet light is effective (cf. Fig. 34, 195, 196). In any event, if there is a strong effect of short-wavelength light and only a slight effect of long-wavelength light (e. g. anthocyanin synthesis in milo seedlings [51]; synthesis of the enzyme cinnamic acid hydroxylase in hypocotyl segments of gherkin seedlings [58]; control of flowering in mustard, Fig. 197) it seems reasonable in the present state of our knowledge to postulate a simultaneous action of phytochrome and of a flavoprotein. A typical response of this sort is discussed in the next section.

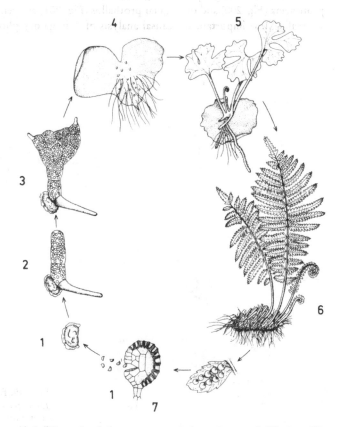

Fig. 198. Representative stages of the life cycle of the common male fern, *Dryopteris filix-mas*. The term "sporeling" designates the germling (2) which originates from a haploid spore (1). Under normal light conditions (that is, white light of at least medium illuminance) the protonema stage (2) is only transient and rapidly followed by a biplanar growth mode (3). The transition from stage (2) to stage (3) can only proceed if a considerable amount of blue light reaches the protonema (after [158])

6. Photomorphogenesis in Fern Gametophytes

a) Basic Phenomena

This section will be restricted to the young gametophytes (= sporelings) of the common male fern, *Dryopteris filix-mas* (Fig. 198). The results which have been elaborated with these sporelings [160a, 161] can be regarded as being representative for the sporelings of many leptosporangiate ferns. The gonospores of *Dryopteris* germinate only in the light. This is a typical phytochrome response [157]. The sporelings, however, can develop "normally" from the very transient filamentous one-dimensional stage, called the "protonema", to the two- or three-dimensional stage, called the "prothallus" only if they receive enough short-wavelength visible light below 500 nm. Under long-wavelength visible light, e. g. red or far-red light of low or higher intensities, these sporelings may continue to grow for many days as cellular filaments which are similar to the filaments of the dark controls (Fig. 199). The characteristic differences in morphogenesis will remain if the culture is continued over a longer period (Fig. 200, 201). If we remember that the highly branched protonema (Fig. 200) and the regular prothallus (Fig. 201) are genetically virtually identical we realize how important the causal analysis of "obligatory photomorphogenesis" might

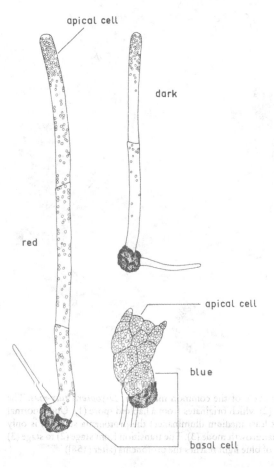

Fig. 199. Representative sporelings of *Dryopteris filix-mas* grown in darkness, continuous standard red light, or blue light. The sporelings were grown for 6 days after germination on inorganic nutrient solution. The blue-grown and the red-grown sporelings have about the same dry weight (after [173])

be for a better understanding of differentiation and development. The problem is to understand in "molecular" terms how blue light can cause the sporelings to switch over from filamentous to two- or three-dimensional growth.

b) The Problem of the Photoreceptor

A good gauge of the photomorphogenic light effect on the sporelings has been the "morphogenic index", i.e. the length of the protonema – or prothallus – divided by its maximal width. This index can be applied because the morphogenic effect of the blue light, i.e. the induction of two-dimensional growth, expresses itself in a very strong inhibition of

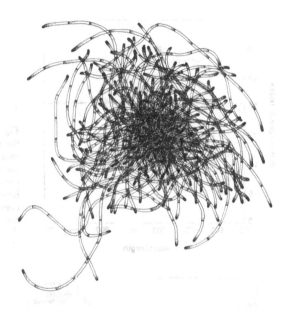

Fig. 200. A manyfold branched fern protonema (*Dryopteris filix-mas*) grown on inorganic liquid medium under continuous standard red light, about 2¹/₂ months after spore germination. The protonema originated from a single spore (after [161])

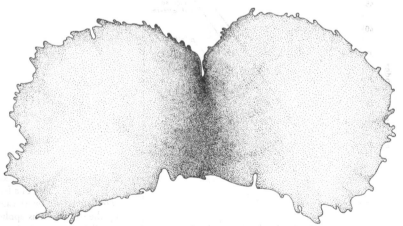

Fig. 201. A typical prothallus of the fern *Dryopteris filix-mas* grown on inorganic liquid medium under continuous standard blue light, about 2 months after spore germination (after [161])

the lengthening of the protonemal stage. A low morphogenic index is the result of a
high morphogenic effectiveness of the light. The action spectrum (Fig. 202) – morphogenic
index as a function of wavelength – shows that the morphogenic effectiveness is very high
below 500 nm. Additional structures appear in the red – a slight peak of action – and
in the far-red, a dip of action. These latter structures can be attributed to phytochrome.
When the sporelings are irradiated simultaneously with red and far-red light at suitable
quantum flux densities the slight morphogenic effect of the red light (lowering of the L/W-
index) is cancelled by the simultaneous far-red light (Fig. 203). On the other hand the
strong morphogenic effect of the blue light does not show any significant interaction with

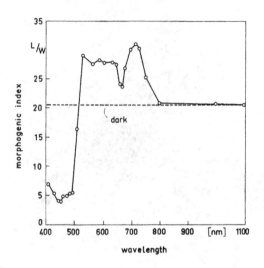

Fig. 202. The morphogenic index L/W of
the protonema of the fern *Dryopteris filix-
mas* as a function of wavelength. Measurements were made after 6 days' culture in
continuous monochromatic light at an
irradiance of 20 µW · cm⁻². Under these
conditions and within this period of time
the sporelings remained filamentous
throughout, even under blue light (after
[158])

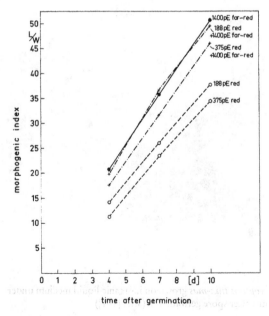

Fig. 203. The effect of red light on the
morphogenic index of a fern protonema
(*Dryopteris filix-mas*) can be nullified
by the simultaneous application of far-
red light at a suitable quantum flux
density (pE stands for pico-Einstein ·
cm⁻² · s⁻¹) (after [266])

far-red light (Fig. 204). All available data indicate that the effect of blue light is due to a photoreceptor other than phytochrome. The photoreceptor involved is possibly a flavo-protein. This conclusion is supported by, among other data, the detailed action spectrum of polarotropism in this system (cf. Fig. 34). The action spectrum of polarotropism in *Dryopteris* is very similar below 500 nm to the action spectrum of polarotropism in *Sphaero-carpus*.

c) A Hypothesis

How can blue light – absorbed by a flavoprotein – initiate two-dimensional morphogenesis? The following hypothesis seems to be appropriate: A blue-light-dependent photoreaction eventually leads to an activation of those genes whose function is required for the "normal", several-dimensional morphogenesis of the fern gametophytes. It is thus assumed that blue-light-specific morphogenesis requires the action of some proteins (enzymes) which are not available or not sufficiently available in red-grown sporelings.

d) Test to Validate this Hypothesis

At the beginning of this section it must be emphasized that all biochemical analyses (partly summarized in [161]) were performed under rigorously controlled standard conditions of growth, i.e. the cultures grew logarithmically on inorganic liquid medium with nearly the same rate of photosynthetic dry matter accumulation in red and blue light. A differential influence of photosynthesis on morphogenesis under blue or red light can probably be excluded in this way.

In an early attempt to validate this hypothesis it was found that blue light strongly increases the protein contents of the sporelings. The decrease of the morphogenic index was found to be inversely correlated with the increase of the relative protein content. Detailed kinetics of the "molecular" and the morphological responses mediated by blue light have shown, however, that the morphological response (Fig. 205) is much faster than

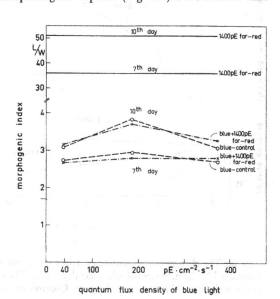

Fig. 204. The effect of blue light on the morphogenic index of a fern protonema (*Dryopteris filix-mas*) cannot be modified significantly by the simultaneous application of far-red light (after [266])

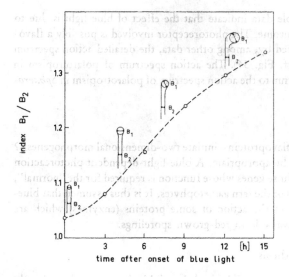

Fig. 205. The increase of the index B_1/B_2 under the influence of blue light. The sporelings of *Dryopteris filix-mas* were grown for 6 days in the standard red light and then placed under standard blue light. The principle of the determination of the index B_1/B_2 is indicated. In the subapical range the influence of blue light on the width of the cellular filament is negligible, that is, B_2 is virtually constant (after [55])

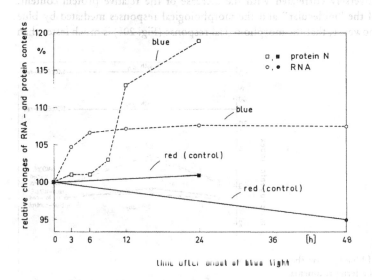

Fig. 206. The changes of the relative protein content and of the relative RNA content under the influence of blue light. System of reference: dry matter. The *Dryopteris* sporelings were grown for 6 days under red light and then placed in blue light. Controls remained in red light (after [55])

the strong increase of the protein content (Fig. 206). The changes of the index B_1/B_2 can be safely measured 3 h after the onset of blue light. Under the same conditions it takes about 9 h before the protein contents increase considerably above the level which is maintained in the red light. At least most of the additional protein which is accumulated under the influence of blue light cannot be regarded as being directly connected with morphogenesis. Rather, it seems to be located in the chloroplast [130].

It was concluded that blue light directs many more primary products of photosynthesis into the channel of protein synthesis than does red light. However, quantitative amino acid analyses of the protein of blue- and red-grown sporelings did not reveal any qualitative difference [45]. The contents of all amino acids – obtained by protein hydrolysis – show about the same increase under the influence of blue light. The pools of the free amino acids are much smaller in the blue than in the red light. This fact indicates that the rate of protein synthesis under the red light is not limited by the pool size of the free amino acids. Blue light seems to exert its promotive influence directly in connection with polypeptide synthesis. These conclusions have been supported by results obtained with a different experimental approach.

When $^{14}CO_2$ is applied to red- or blue-grown sporelings it is found that much less ^{14}C is present in free amino acids in the blue-grown as compared to the red-grown sporelings (Fig. 207). On the other hand, much more of the ^{14}C is incorporated into protein in the blue as compared to the red light (Fig. 208). It was concluded that chloroplasts in blue-grown sporelings synthesize protein more rapidly than chloroplasts in red-grown sporelings.

However, more refined experiments indicated that the blue-light-dependent promotion of protein synthesis is probably not restricted to the chloroplast compartment. The capacity for protein synthesis was studied in sporelings which were first grown under red light and then transferred to blue light [198]. The rate of protein synthesis was measured in short-term experiments (40 min) using $^{14}CO_2$. The photosynthetic rate was the same in red and blue light; it was not influenced by the transfer. Likewise the rate of ^{14}C incorporation into the pool of free amino acids was not significantly different in red and blue light. On the other hand, the rate of incorporation of ^{14}C into protein increased rapidly after the transfer of the sporelings from red into blue light (Fig. 209). The same results (no influence of blue light on the specific activity of the free amino acid; a strong promotive influence on the specific activity of protein-bound amino acid) was observed in the case

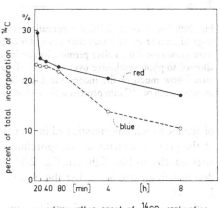

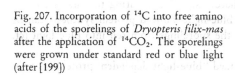

Fig. 207. Incorporation of ^{14}C into free amino acids of the sporelings of *Dryopteris filix-mas* after the application of $^{14}CO_2$. The sporelings were grown under standard red or blue light (after [199])

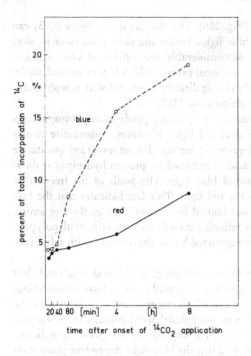

time after onset of $^{14}CO_2$ application

Fig. 208. Incorporation of ^{14}C into protein-bound amino acids of the sporelings of *Dryopteris filix-mas* after the application of $^{14}CO_2$. The sporelings were grown under standard red or blue light (after [199])

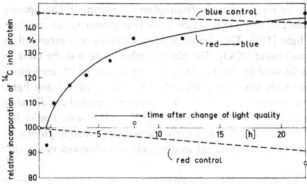

Fig. 209. The increase of ^{14}C incorporation from $^{14}CO_2$ into protein-bound amino acids of the sporelings of *Dryopteris filix-mas* after a transfer of the red-grown sporelings to blue light. The sporelings were grown for 8 days after germination under red light. After the transfer to blue light samples were allowed to photosynthesize over a period of 40 min in $^{14}CO_2$. Immediately afterwards it was determined how much of the incorporated activity was present in protein-bound amino acids (relative incorporation of ^{14}C into protein at the end of the red-light growing period = 100%) (after [198])

of alanine which was investigated in detail [198]. The result was striking: While the increase of the protein contents of the sporelings is hardly significant during the first six hours after transfer to blue light (cf. Fig. 206), the blue-light-induced protein synthesis can be detected about one hour after the transfer of the sporelings from red into blue light (cf. Fig. 209).

It was concluded that this rapidly-synthesized "blue-light-dependent protein" is only a small fraction of the total protein of the sporeling and might be directly related to morphogenesis, while the bulk of the blue-light-dependent protein is chloroplast protein. This latter suggestion was recently confirmed in experiments with isolated chloroplasts [213]. Irradiation of the gametophytes of *Pteridium aquilinum* with blue light led to a nearly 5-fold increase in the amino-acid-incorporating activity of isolated chloroplasts. The gametophytes were grown in red light and then exposed for some time (24 to 72 h) to blue light. Under these conditions a dramatic increase in the amino-acid-incorporating activity of the isolated chloroplasts was consistently observed.

e) Microscopic Data on Plastids

Microscopic observations support the biochemical data. BERGFELD found that the size of the chloroplasts, as measured in the basal cell of the sporelings (cf. Fig. 199), is determined by the light quality. In blue light the chloroplasts are much larger than in red light (Fig. 210). The process of blue-light-dependent plastid growth is reversible. The size of the chloroplasts adjusts at any given time to the value characteristic of the particular light quality (Fig. 211). It is discovered in such experiments that even "mature" plastids are subjected to modulations and that their size and structure can be controlled by photomorphogenic light independently of photosynthesis even in specialized cells which no longer grow.

f) Data on RNA

In *Dryopteris filix-mas* as well as in *Pteridium aquilinum* higher RNA contents of the sporelings were observed under blue light than under red light [55, 211, 212]. In transfer experiments it was found that the relative RNA contents of red-grown sporelings increases rapidly after the transfer to blue light (cf. Fig. 206). The conclusion that the rate of RNA synthesis is increased under these conditions was supported by the observation that there is a rapid blue-light-dependent increase of the specific activity of the total RNA after the application of ^{14}C-uridine.

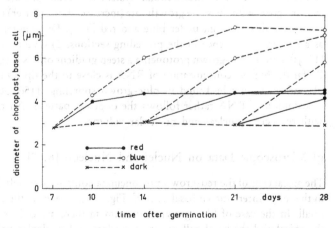

Fig. 210. Control by light of the size of the chloroplasts in non-growing basal cells (cf. Fig. 199) of the sporelings of *Dryopteris filix-mas*. Obviously the size of the chloroplasts is determined by the quality of light, standard red or blue light (after [5])

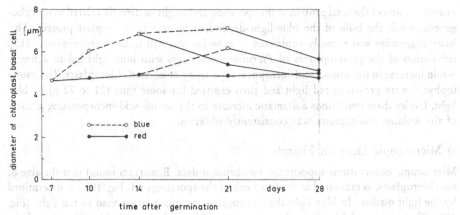

Fig. 211. Control by blue light of the size of the chloroplasts in non-growing basal cells of red-light-grown sporelings of *Dryopteris filix-mas*. After the transfer of the cultures to blue light the chloroplasts increase in size. This increase is reversible (after [5])

However, the importance of differential RNA synthesis for blue-light-dependent morphogenesis is not yet clear [30]. At least two factors make a satisfactory interpretation of the biochemical data difficult: 1) The RNA content per average cell decreases under all experimental conditions in the course of development, whereby the RNA *per cell* is about 40 per cent higher under red than under blue light [56, 30]. Obviously the problem of the appropriate "system of reference" comes into force. Blue light specifically increases DNA replication and mitotic activity even if the sporelings have the same growth rate under blue and red light, as determined by dry matter increase: Cell number and DNA content of sporelings of the same age and the same dry matter content are much higher under blue than under red light [56]. It has been pointed out [56] that the "cell" (or "unit DNA") may *not* be used as a system of reference for biochemical data such as protein or RNA contents in the case of the fern sporelings. The only useful system of reference seems to be the entire multicellular system if precautions are taken to ensure that the systems grow with the same growth rate (increase of dry matter) under the different experimental conditions (e.g., under blue and red light). Dry matter has therefore been used as a system of reference in the preceding sections. 2) Cytochemical studies have shown [156] that in the red-grown protonema a steep gradient of RNA exists along the protonema. By far the highest concentration of RNA is close to the tip. On the other hand this type of gradient cannot be found in blue-grown sporelings [156, 30]. Here the longitudinal distribution of RNA rather follows the opposite pattern. An analogous situation exists with respect to nuclear and nucleolar volume.

g) Microscopic Data on Nuclei and Nucleoli [6, 7]

The apical cells of the red-grown protonemata contain extremely large nuclei and nucleoli; in the fully-differentiated basal cell (cf. Fig. 199), however, the nucleus and nucleolus are small. In the case of blue-grown protonemata there are only slight differences between the apical and the basal cell as far as nuclear and nucleolar volumes are concerned. In the apical cell these structures are much smaller in the blue-grown protonema than in the red-grown protonema; in the basal cell, however, the nucleus and nucleoli are larger

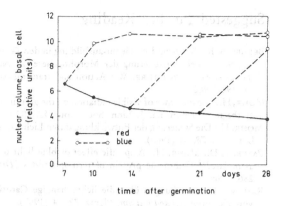

Fig. 212. Control by light quality of the nuclear volume in non-growing basal cells (cf. Fig. 199) of the sporelings of *Dryopteris filix-mas*. The nuclear volume (one unit = 55 μm³) increases up to a certain level if the sporelings are placed under standard blue light instead of standard red light (after [6])

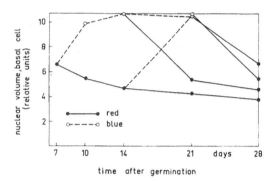

Fig. 213. Control by light quality of the nuclear volume in non-growing basal cells of the sporelings of *Dryopteris filix-mas*. The increase of nuclear volume (one unit = 55 μm³) which occurs under standard blue light is more or less reversible. It is seen that the nuclear volume adjusts to the value characteristic of the particular light quality (standard red or blue light) (after [6])

in the blue-grown sporeling [6, 7]. Since the nuclei in the basal cells no longer divide, the question of whether or not they still respond to changes of the light conditions was of great interest. This is indeed the case. Fig. 212 shows that the volume of the nuclei increases up to a certain level when the sporelings are placed under blue light instead of red light. The same is true for the nucleoli. Fig. 213 shows that these shifts in nuclear volume are reversible: if the sporelings are kept under red light instead of blue the nuclear volume decreases to the level which is characteristic of the controls which have only received red light. It seems that the change of nuclear and nucleolar volume reflects a change of nuclear function. The increase of volume in the blue light could be a manifestation of a higher rate of nuclear activity under these conditions.

h) Summary

Normal morphogenesis of fern gametophytes (= sporelings) is blue-light-dependent. The photoreceptor involved is probably a flavoprotein. A distinction must be made between two distinct effects of the blue light. Firstly, blue light dramatically increases the protein-synthesizing capacity of the chloroplasts. Secondly, blue light causes, presumably through differential gene activation, synthesis of the enzymes which are required for initiation and maintenance of two-dimensional (biplanar) morphogenesis. Unfortunately, it has not been possible so far to detect these enzymes directly.

Suggested Further Reading

BERGFELD, R.: Kern- und Nucleolusausbildung in den Gametophytenzellen von *Dryopteris filix-mas* (L.) Schott bei Umsteuerung der Morphogenese. Z. Naturforsch. *22b*, 972 (1967).

DELBRÜCK, M., SHROPSHIRE, W.: Action and transmission spectra of *Phycomyces*. Plant Physiol. *35*, 194 (1960).

MOHR, H.: The influence of visible radiation on the germination of archegoniate spores and the growth of the fern protonema. J. Linn. Soc. (Bot.) *58*, 287 (1963).

MOHR, H.: Die Steuerung der Entwicklung durch Licht am Beispiel der Farngametophyten. Ber. dtsch. bot. Ges. *78*, 54 (1965).

PAYER, H.D., MOHR, H.: A specific effect of blue light on the incorporation of photosynthetically assimilated ^{14}C into the protein of fern sporelings (*Dryopteris filix-mas* (L.) Schott). Planta *86*, 286 (1969).

RAU, W.: Untersuchungen über die lichtabhängige Carotinoidsynthese. I. Das Wirkungsspektrum von *Fusarium aquaeductuum*. Planta *72*, 14 (1967).

The Problem of Phototropism

1. The Traditional View [76]

It is stated in every textbook of botany that the direction of growth of stems, roots and leaves is influenced by light as well as by gravity. In this phenomenon of plant growth alterations by unilateral light, referred to as phototropism, the stems are generally positively phototropic, roots generally negatively phototropic (if they respond at all), and leaves plagiotropic (Fig. 214). It is further stated that "in both photo- and geotropism the curvature is the result of differential growth on the two sides of the plant axis. The side towards which the curvature occurs grows less rapidly than the side opposite and the resultant of this differential growth is the curvature" [76].

While the analysis of the phototropic response has led to the discovery of auxin, the "mechanism" of the phototropic response is still a matter of debate. Among the explanations which have been advanced are asymmetry in auxin production at the very apex, lateral transport of symmetrically-produced auxin, asymmetric destruction of auxin and asymmetrically-inhibited longitudinal transport of auxin [76]. While there seems to be agreement now that a lateral transport of indoleacetic acid occurs under conditions of tropistic stimulation, the mechanism through which a unilateral stimulus of light becomes transduced into lateral auxin transport is still obscure [44]. The established concept of phototropism (as quoted above) requires rigorous refinement in order to be consistent with the facts. We shall concentrate on two points.

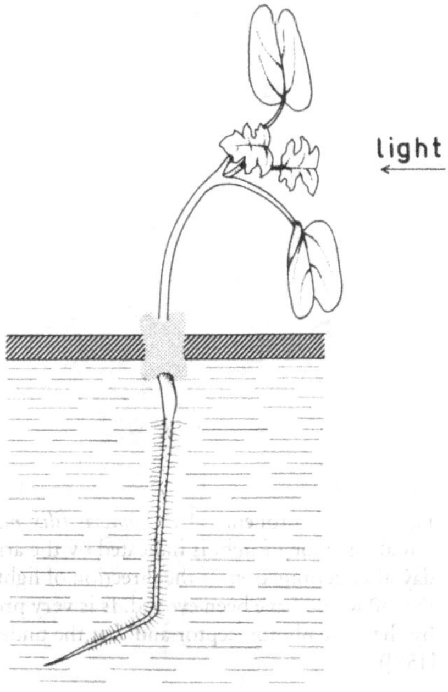

light

Fig. 214. Typical growth alterations of plant organs under the influence of unilateral light (arrow). The hypocotyl of the mustard plant responds positively and the taproot responds negatively, while the cotyledons and the consecutive leaves express plagiotropism (after [22])

2. Phototropic Responses in a Unicellular System
(Example: Filamentous Fern Sporelings ("Chloronema"; cf. Fig. 199))

A filamentous fern sporeling (*Dryopteris filix-mas*) will readily respond with a distinct phototropic curvature if the direction of light is changed (Fig. 215). However, the curvature is *not* the result of a differential growth on the two sides of the apical cell. Rather, the curvature results from a shift of the "growing point" of the cell from the center of the apex to the flank (cf. Fig. 31). Growth on the shaded flank is not significantly changed. Thus, the "mechanism" of the phototropic (and polarotropic) response cannot be understood in terms of the established concept ("the side towards which the curvature occurs grows less rapidly than the side opposite and the resultant of this differential growth is the curvature"). The action spectrum of the photo- and polarotropic response of the fern chloronema (cf. Fig. 34) shows a double peak. The action of the red range ($\lambda > 520$ nm) can be attributed to phytochrome whereas at least most of the strong action of blue and near UV light might be mediated by a flavoprotein [244]. We recall that in sporelings

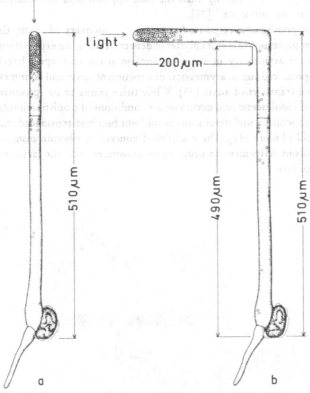

Fig. 215. A protonema of *Dryopteris filix-mas* 6 days (a) and 7 days (b) after spore germination. The actual direction of light is indicated by the arrows. The direction of light was turned by 90° at the fifth day after germination. If the direction of light had been turned by 50° only, the response as pictured in Fig. 30 would have been evoked. Is is very probable that polarotropism and phototropism are mediated by the same photoreceptor and that the underlying "mechanisms" of the responses are the same (after [158])

of *Sphaerocarpos donnellii* (a liverwort) the photo- and polarotropic response can only be evoked by blue light (cf. Fig. 34). In this organism phytochrome does not seem to be related to the phototropic (and polarotropic) response.

3. Phototropic Responses of the Hypocotyl
(Example: Seedling of Mustard, *Sinapis alba;* cf. Fig. 3)

In this multicellular system the phototropic curvature is indeed the result of differential growth on the two sides of the organ. Action spectra clearly indicate that only short-wavelength visible light can elicit a phototropic response. The action spectra obtained with different systems (including the grass coleoptile) are very similar. The action spectrum obtained for the photo- and polarotropic response in *Sphaerocarpos* (cf. Fig. 34) and the action spectrum elaborated for the light-growth response in sporangiophores of *Phycomyces* (which is identical with the action spectrum for phototropism in this system) may serve as examples (cf. Fig. 196). It is very probable that this type of action spectrum reflects the involvement of a flavoprotein (or flavoproteins) as photoreceptors. As already pointed out, at present there is no satisfactory model for the mechanism by which a multicellular system evaluates an asymmetric, azimuthal gradient of light intensity and then transduces this information into altered patterns of elongation.

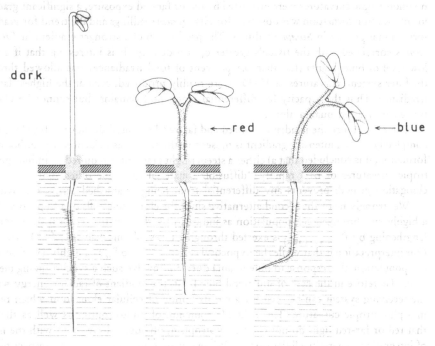

Fig. 216. These drawings illustrate the observation that unilateral red light ($\lambda > 550$ nm) does not cause any phototropic curvature in the mustard or buckwheat hypocotyl although longitudinal growth is strongly (but not totally) inhibited and a gradient of light intensity probably exists in the hypocotyl even for red light. On the other hand, the phototropic sensitivity towards blue light ($\lambda > 500$ nm) is very high under all circumstances (after [174])

While a quantitative theory of blue-light-mediated phototropism is still in its infancy, a particularly troublesome situation with regard to higher plants exists in the red and far-red part of the spectrum. It is well-known that red and far-red light inhibit the elongation of the hypocotyl in higher plants (cf. 10th lecture). It is also well-known that this inhibition in *Sinapis* or *Lactuca* seedlings is due solely to inhibition of cell elongation [88]. Thirdly, it is also well-established that unilaterally-given red or far-red light do not produce phototropic curvatures, even in dosage ranges where elongation has not been totally inhibited (Fig. 216) [174], that is, the possibility for differential changes in elongation exists if a gradient in light intensity can be set up across the tissue. While some authors (e.g. [73]) have explained this behavior by stating that hypocotyl tissue is relatively transparent to red and far-red light, more recent experiments indicate that this explanation is not satisfactory. Experiments performed with mustard (*Sinapis alba*) and buckwheat (*Fagopyrum esculentum*) seedlings may serve as an example [237]: No phototropic curvatures were observed for unilateral exposures at any irradiance of red or far-red light. These irradiances were in the range where cell elongation in the hypocotyl is inhibited but never totally inhibited. For the production of a curvature greater than 30° a difference in elongation on opposite sides of the hypocotyl of only 3 per cent is needed. Thus, the potential for differential elongation to produce significant curvatures was present at every irradiance used. In addition anthocyanin synthesis was measured in hypocotyl halves excised from *Fagopyrum* and *Sinapis* seedlings irradiated unilaterally or equilaterally with red or far-red light. Although no phototropic curvatures were produced by red or far-red exposure, a significant gradient in anthocyanin formation was observed for *Fagopyrum* seedlings and the trend for gradient formation is present in *Sinapis* seedlings. The production of a stronger gradient in *Fagopyrum* is correlated with the tissue's greater optical density. It is interesting that if a very low level of blue quanta (less than 0.1 per cent of total irradiance) was allowed through the filter system, curvatures of 15-20° were readily produced, even at the highest far-red irradiance. Thus the capacity for differential elongation remains, but cannot be elicited by red or far-red quanta alone.

Therefore, since the irradiances of red and far-red light used do not inhibit elongation completely and an intensity gradient is present in the tissue, as evidenced by anthocyanin formation, it is concluded that a) either a steeper light gradient is required to induce phototropic curvatures or b) a rapidly "diffusible" material (or some other signal) affecting elongation growth prevents any differential from being established across the tissue.

We strongly favor the second alternative since the elongating cells of a hypocotyl form a highly-synchronized cell population as indicated by the fact that control of hypocotyl lengthening by $P_{fr\ (ground\ state)}$ is exerted through a threshold mechanism (cf. 10th lecture). Our interpretation is that while the hypocotyl cells respond to P_{fr} as a highly-synchronized cell population, the response to blue quanta does not involve such a synchronizing mechanism. Therefore in the case of unilateral blue light, the gradient of radiant energy within the receiving system can be used for a redistribution of cellular elongation which results in a phototropic curvature. Thus, the phenomenon of phototropism as well as the fact that red or far-red light do not elicit phototropism, obviously have to do with the mode of integration in a multicellular system. While our present knowledge with respect to this point is very limited, the problem at least may be outlined briefly.

4. Homeostasis and Integration

W.B. CANNON formulated the concept of homeostasis according to which a body acts to maintain a stable internal environment through the interaction of various physiological processes. Thus, coordinated physiological processes will maintain most of the steady states in the organism. Our present-day knowledge suggests that this concept can be extended to coordination during development, that is, while irreversible temporal changes of the system occur, including *irreversible* changes of steady states (cf. Fig. 143). The coordinating processes which occur during development may be called "integration". Every organism responds as a whole, i.e. as a functional unit. To our present knowledge integration within this functional unit can be exerted through humoral and nervous mechanisms. In the plant body only humoral mechanisms are believed to bring about integration. However, our present knowledge about plant hormones is by no means sufficient to explain the phenomena of integration in the multicellular plant system. The integration phenomena may be classified as:

1. Integration within the cell (e.g. between compartments or organelles)
2. Integration between cells
3. Integration between tissues
4. Integration between organs

In many of the traditional biochemical experiments the aspects of integration have been deliberately neglected. However, for the physiologist "integration" is one of the most important aspects and one which can only be approached by studying the intact system while keeping in mind, of course, the basic knowledge elaborated by biochemistry about the constituents and elementary functions of the cell and of subcellular organelles. Many of the most fruitful studies in science have been made on tissue extracts, homogenates, and in isolated particulate fractions. However, after completion of these analytical studies with the methods of biochemistry, the physiological problem is to investigate the part they play in the functioning of the organism, their integration with other functions, their homeostasis, and their regulation in the course of *development* of the integrated multicellular system.

Three phenomena which were described in the course of the foregoing lectures require a conception of integration considerably more sophisticated than the vague humoral concepts used up till now in plant physiology. These phenomena are: threshold regulation of lipoxygenase synthesis (7th lecture); threshold regulation of longitudinal growth (10th lecture); lack of phototropism under unilateral red or far-red light (this lecture). It is felt that the occurrence of plasmodesmata, that is, the existence of a symplasm, might be the structural basis for the very rapid and efficient communication between plant cells in a multicellular organ. However, the *means* of communication are unknown.

Suggested Further Reading

BRIGGS, W.R.: Phototropism in higher plants. In: Photophysiology I (A.C. GIESE, edit.) New York: Academic Press, 1964.

GALSTON, A.W., DAVIES, P.J.: Control Mechanisms in Plant Development. Englewood Cliffs: Prentice-Hall, 1970.

SHROPSHIRE, W., MOHR, H.: Gradient formation of anthocyanin in seedlings of *Fagopyrum* and *Sinapis* unilaterally exposed to red and far-red light. Photochem. Photobiol. *12*, 145 (1970).

Genes and Environment

1. The Problem

It is known that the basic laws of classical genetics were discovered using higher plants. Both Mendel and Correns worked with sporophytes of higher plants (*Pisum sativum*, *Mirabilis jalapa*) when they discovered (or re-discovered) the so-called Mendelian Laws. More recently, the main body of information about molecular genetics was predominantly obtained with phages, bacteria and molds. The laws which were discovered with these so-called lower organisms have been extrapolated to higher multicellular systems, including man. No scientist will doubt that such an extrapolation is justified in principle.

One of the basic problems of present-day biology is to understand those laws which govern the interaction of the genome and the environment of a living system. In order to comprehend the full importance of this phenomenon we have to remember that the specific development of every living system depends on the genetic information of the particular system and on its environment. The most important factor of the environment is light, at least in all higher plants. It is important, however, to realize that even light does not carry any specific information. Light, like every other environmental factor, can only be regarded as an *elective* factor which influences the manner in which the genes which are present in the particular organism are used.

Higher plants seem to be the most useful living systems for the scientific study of the interaction between genome and environment. Accordingly, control of development by light in the sporophytes of higher plants was chosen as a model system by a number of investigators (including two masters of classical plant physiology, PFEFFER and KLEBS) to determine the principal laws which govern the interaction of genes and environment in the life cycle or ontogeny of a higher multicellular system, including man.

This research topic, which we call "developmental genetics and environment", ist not just a scientific matter. Obviously there are great practical and even political implications. For instance, we must know the physiological laws which govern the interaction of genes and environment in order to improve (and eventually optimize) our educational system. Many sociologists and even some anthropologists assume that differences in behavior among human populations are determined almost entirely by cultural variables, and that the genetic contribution to differences in brain function and behavior is at best trivial (e.g. [177]). While such a conclusion is not compatible with recent experimental data (e.g. [126, 78, 297, 300]) it is becoming extremely difficult to study the problem of "genes and environment" in a scientific manner, that is, relatively free of pejorative or ideological implications. It seems wise, therefore, to concentrate on higher plants (as our scientific predecessors did in the case of classical genetics) to formulate the principal laws which govern the interaction of genes and environment in higher systems and to check the validity of these laws for man after the laws themselves have been firmly established.

2. The Principal Result of the Foregoing Lectures

The two potato plants in Fig. 2 are genetically identical. The seemingly dramatic difference between these two systems is due to the fact that one plant grew in the light, and the other in the dark. The phenomenon of photomorphogenesis, that is the control of characteristics (or traits) by light, is clear. It is very probable that the *difference* between the two plants must be attributed exclusively to the active phytochrome, P_{fr}, which has been formed in the light-grown plant whereas in the dark-grown plant only P_r, the physiologically inactive species of the phytochrome system, is present in the cells.

3. Environmental Variability of Different Characteristics (or Traits)

When we follow the development of a potato plant (cf. Fig. 2) or of any sporophyte of a higher plant, we notice that the vegetative parts of the plant (including stems, leaves and inflorescences) are much more variable as far as the influence of light is concerned

Fig. 217. The differences between these three specimens of the species *Gentiana campestris* are determined by the varying altitude of the habitat (after [141])

than the reproductive parts. The characteristics of the flowers (including stamens, pistils, corolla and calyx) are hardly modified by light. A potato plant will form very similar flowers under all circumstances which allow flower formation at all. Since this is true for all flowering plants (Fig. 217), the taxonomy of spermatophytes has been based predominantly on the characteristics of the reproductive parts.

While the characteristics of the flowers are hardly modified by the environment, they are very susceptible to genetic changes. Fig. 218 shows the inflorescence of *Linaria vulgaris*,

Fig. 218. Inflorescences of *Linaria vulgaris*. Left, a normal specimen with dorsiventral flowers; right, a mutant with radially symmetrical flowers (after [176a])

on the left from the normal form, on the right from a mutant. The plant on the left will produce the same dorsiventral flowers under all circumstances which allow flowering; the mutant on the right will produce radially symmetrical flowers under all conditions which allow flowering.

In conclusion, there is good evidence [4] for the view that in general the dimensions of stems, leaves and inflorescences, that is, the dimensions of the vegetative parts, have a much greater range of environmental variability than the reproductive parts of the same plant.

4. A Hypothesis to Explain the Phenomena Described in the Previous Section [176a]

The main feature of the hypothesis is that complicated polygenic characteristics as represented by the reproductive parts of a flowering plant show a very limited environmental variance in general because *internal* control over the development of these complicated polygenic characteristics can be exerted simultaneously at several or even many points of the causal network ("developmental homeostasis" [4]). On the other hand, vegetative traits such as lengthening of an internode, growth of a leaf, synthesis of chlorophyll or anthocyanin in a seedling, show a high degree of environmental variance. These characteristics have been called "open characteristics" since their manifestation (or quantitative expression) is thought to be controlled at one or at a few points of the causal network by external stimuli such as light via P_{fr}. Several examples of "open characteristics" have been described in the previous lectures (cf. Figs. 91, 127, 159).

If the validity of this hypothesis could be fully established for all higher systems (including mammals and man), a prediction derived from this hypothesis would be that the quantitative expression of those polygenic traits which are bound to the central nervous system such as electroencephalographic pattern, intelligence quotient, and temperament, can be influenced by the environment only to a very limited extent.

This concept is supported by evolutionary arguments. A flower which locates its pollen on the insect's body incorrectly is likely to have no progeny [4]; a man who did not think clearly and consistently was likely to have no progeny in the course of hominid evolution. (It is realized, of course, that the situation has changed in modern societies).

5. Phenotypization of Genetic Information as a Two-step Process

In most cases the real situation is more complex than has been described so far. It must be realized that the influence of the environment on the development of a characteristic (or trait) can be at least two-fold. 1) The environment can determine (e.g. through a threshold response) whether or not a characteristic can be expressed at all. 2) The environment can modify the quantitative expression of the characteristic after the requirements of the first step are fulfilled.

As an example we can recall the phenomenon of flowering in a short-day plant, *Kalanchoe blossfeldiana* (cf. Fig. 13). This plant is an *obligatory* short-day plant, and the decision about flowering or non-flowering can be made by P_{fr} which is absent or present above a certain threshold value in the middle of the long night (cf. Fig. 187). Fig. 13 illustrates this all-or-none response: 8 h white fluorescent light per day leads to flowering; 8 h white fluorescent light + 1 min red light in the middle of the dark period result in a vegetative plant; a program of 8 h white fluorescent light plus [1 min red + 1 min far-red light] in the middle of the night results in flowering. The second important point is that light does not have any significant influence on the development of the flower once the induction of flower initiation has occurred. Under all circumstances which permit flowering, a *Kalanchoe* plant will form virtually the same flowers.

I suggest that the development of many characteristics in man is controlled in a similar way. The IQ problem is probably a good example. The normal intelligence of a human being as measured by the IQ is a polygenic characteristic which depends on an intact central nervous system. If a child undergoes chronic protein malnutrition (or other deficiencies) during the early years of its life, the development of the central nervous system can be irreversibly hindered, and correspondingly an IQ within the range of normality cannot develop. Stated in general terms, physiological disadvantage in prenatal and postnatal development can substantially lower phenotypic IQ and thus reduce the genotype-phenotype correlation. However, if nutrition and other environmental factors (including care-taking and stimulation) allow the normal development of the central nervous system, it seems that the actual value of the IQ can only be influenced slightly by the environment. The inheritability of intelligence in white, middle-class populations of schoolchildren and adults has been repeatedly estimated to account for 60 to 80 per cent of the total variance in general intelligence scores, in whatever way they were measured [31, 122]. Recent investigations on early mental development in man have led to the finding [297] that infant mental development is primarily determined by the genetic blueprint and that, except in unusual (that is, pathological) cases, other factors served mainly a supportive function.

6. Future Topics for Research

The problem of how much genetic and environmental variances contribute to the total variance of a given characteristic in a population must be investigated by scientific means for a great many characteristics, in plants, animals and in man. This is an experimental as well as a theoretical task.

1. Many more experiments with higher systems are needed to investigate quantitatively a multiplicity of characteristics (including biochemical and "molecular" traits) with respect to the contribution of genetic and environmental variances to the total variance of the characteristic in the population. This type of experiment requires the availability of facilities for controlling every important parameter of the environment ("phytotrones"). In the past such experiments have often been done in agricultural research but considerable improvement is needed in the precision and the logics of the experiments. In principle the experiments follow a simple pattern [233]: Draw two random samples of seeds from the same genetically heterogeneous population. Plant one sample in uniformly good conditions, the other in uniformly poor conditions. The average difference (with respect to an explicitly defined characteristic) between the populations of plants will be entirely environmental (cf. Fig. 1), although the individual differences in the particular characteristic within each sample will be predominantly genetic (Fig. 219). With known genotypes of seeds (or spores, tubers (cf. Fig. 2), cuttings, etc.) and precisely defined environments, genetic and environmental variances between groups can be studied with respect to different characteristics. The hypothesis has been advanced that in the case of polygenic characteristics the environmental variances are small. It now becomes an experimental task to validate this hypothesis for higher systems in general.

2. A theoretical approach (preferably in terms of general system theory) must be undertaken to check the so far vague hypothesis that complicated polygenic characteristics are generally characterized by developmental homeostasis and thus necessarily (that is, for

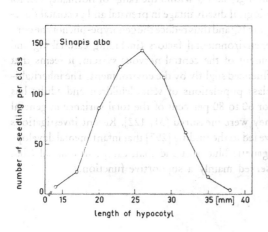

Fig. 219. Frequency distribution of a population of mustard seedlings with respect to length of hypocotyl at 72 h after sowing. The distribution curve is very similar to a normal distribution. Since all seedlings are exposed to the same environmental conditions [cf. 162] the variance of the characteristic must be due to genetic variance (after [221]). The frequency distribution of the trait remains approximately the same even if the means differ by a factor of 8 (cf. Fig. 65). While the average hypocotyl length difference between the populations of seedlings (cf. Fig. 1) are entirely environmental, the individual differences in hypocotyl length within each sample are entirely genetic. It is obvious that the relative contribution of genetic and environmental variances to the total variance can be estimated precisely in this type of experimental approach

theoretical reasons) show a relatively small environmental variability. The latter means that the variance of the characteristic in a given population is predominantly genetic variance.

As pointed out in a previous section of this lecture, this type of scientific research has considerable practical and political implications. For the sake of our society, in particular for the sake of our children, responsible politicians must be prepared to accept scientific truth in this field even if the scientific argument contradicts the fashionable ideological belief.

"The greater the environmental equality, the greater the hereditary differences between levels in the social structure. The thesis of egalitarianism surely leads to its antithesis in a way that KARL MARX never anticipated" [233].

"To the extent that better, more supportive environments can be provided for all children, genetic variance and mean scores will *increase* for all groups. Contrary to the views of many naive environmentalists, equality of opportunity leads to bigger and better genotype-phenotype correlations. It is toward this goal that socially concerned citizens should work" [234].

This brief discussion of the field of study of "genes and environment" which is at present characterized by the conflict of ideological prejudice with scientific approach leads us directly to the theme of the concluding lecture, which will be concerned in general terms with the responsibility of the knowledgeable and experienced scientist in the modern world.

Selected Further Reading

BERG, R.L.: A general evolutionary principle underlying the origin of developmental homeostasis. Amer. Naturalist *43*, 103 (1959).

Environment, Heredity, and Intelligence. Compiled from the Harvard Educational Review. Reprint Series No. 2. Harvard Educational Review, Cambridge, Mass., 1969.

MOHR, H., SITTE, P.. Molekulare Grundlagen der Entwicklung (chapter 23.1). Munchen: BLV, 1971.

SCARR-SALAPATEK, S.: Race, social class, and IQ. Science *174*, 1285 (1971).

Epilogue: Science and Responsibility[1]

The last time I talked to this audience[2] I gave a scientific lecture, that is, I presented a hypothesis and cases which supported the hypothesis, and I pointed out that there are no data known at present which do not agree with the hypothesis. In other words, I tried to follow carefully the established manner of dealing with a scientific problem. Today I shall not discuss a scientific problem, but I wish to deal with a certain aspect of science itself. When a scientist tries to contribute to the discussion on this level he finds himself in a difficult position, for it is hardly possible to deal with the problem of science in precisely the same way that we deal with a scientific problem. To say it frankly and in advance: Any philosophy of science is less rational than science itself.

One of the main problems in this connection has been that we traditionally use verbal language when we talk about science. Every verbal language contains a great number of semantic overtones. While such semantic overtones are the very essence of poetry and literary prose and the stock-in-trade of the political orator, verbal language is an awkward tool for the scientist who is trying to express himself without also evoking an emotional response. The words have all been used before, often in emotional situations, and the word recalls the emotion whether one wants it to or not [286].

Science can be defined operationally as a venture of the human mind which aims at genuine knowledge. Genuine scientific knowledge is expressed as true propositions: individual propositions (or data or facts) and general propositions (or laws). True general propositions are those which describe the behavior of scientific models which are satisfying, and thus they refer to the behavior of real systems: living or non-living systems; those existing in nature and those made by man. The criterion of truth with respect to scientific propositions is prediction which is correct when compared with subsequent experience.

My thesis is that the responsibility of the scientist, the responsibility of the scientific community, is to guarantee the truth (and therewith the reliability) of scientific propositions. This, of course, has been the traditional responsibility of the scientific community.

Every experienced member of the scientific community knows the difference between true and false propositions exactly, and every experienced scientist knows how easy it is to make statements and how difficult it is to prove that the propositions are true. However, I would like at least to give precision to the term 'law' in this context: the highest degree of scientific truth is reached when laws can be formulated, universal or particular laws. Universal, if they apply to every real system, and particular, if their validity is restricted to a certain class of systems. The first and second laws of thermodynamics are famous examples of universal laws; on the other hand, the proposition $\Delta G \neq 0$ is a thermodynamic

---

1 This lecture was delivered to a general audience.
2 This phrase refers to the 15th lecture ("Light-mediated flavonoid synthesis: a biochemical model system of differentiation") which was also delivered to a general audience.

law which is only applicable to open systems, including living systems. For living systems, $\Delta G \neq 0$ is a universal law, and so on.

The second introductory remark I would like to make concerns the significance of authority in science. In principle, the scientist recognizes no authority except an empirical observation of nature. In practice he will often accept a report of such an observation by a single established or distinguished researcher but in general a consensus of several researchers is preferred. In any case, for the scientist empirical observation is the ultimate court of appeal which must be invoked if there is even the slightest doubt with regard to any statement regarding his science. The true propositions of science do not depend on any political authority or ideology. Scientific thinking can be suppressed by crude political power, as, for example, during the Stalin era in Russia, but as long as science is permitted in a society, it is independent of the particular political system or ideology.

Now let me return to my previous thesis, that the goal of science is genuine knowledge about real systems. The responsibility of a scientist in relation to science follows from this definition of the goal of science: the moral obligation of a scientist is to contribute to the wealth of genuine knowledge under all circumstances, even under economic or political pressure, and by doing so to decrease the amount of prejudice and superstition about the real systems of nature, including man.

Two problems will arise immediately. The first problem is that every scientist is a member of the world-wide scientific community and at the same time a citizen of a country, that is, an element in a political system. The latent problem becomes crucial when moral loyalty to the scientific community and loyalty to the political system come into collision. This conflict of loyalties arises automatically if absolute intellectual freedom, an indispensable prerequisite of scientific thinking, is no longer guaranteed in a political system. This is the case in every country where a single political ideology, a specific political dogma, rules the scene, as for instance in those countries ruled by dogmatic marxism. However (and this brings me to the second problem), in the modern world technology is almost exclusively based on individual or general scientific propositions, or, in simpler terms, on genuine scientific knowledge. No country can afford to base technology on political dogma, ideology, agnostic superstition or prejudice, even for a limited time. Modern technology can only be based on genuine knowledge. This, of course, is the main reason why at least some scientific research is supported by every country whose existence requires a sophisticated technology. And this is the reason why no political power can afford to suppress the freedom of thinking in scientific research, even if most of the country's activities are controlled by a rigid ideology. Each time I have visited Russia or East Germany I have felt that our fellow scientists there live in a sort of natural park provided with an impenetrable fence. Inside the fence there is full intellectual freedom, but outside the fence there is none.

The relationship of science and technology has recently been confused by a number of misconceptions. I hope to avoid these difficulties by operationally defining science and technology. Let me first repeat my operational definition of science: the goal of science is to gain genuine knowledge about the real world, in brief, to understand the real world. On the other hand, the goal of every technology is to change or manipulate the real world for the sake of man, or to be more precise, for the sake of some men. Both aspects are always present in a particular research project in science or technology even though the researcher is concerned with only one aspect. Every true scientific proposition can be used technologically; and every technological research may contribute to the progress of science. Einstein was directly concerned with basic scientific research when he postulated and tested the model leading to the equation $E = m \cdot c^2$; yet this same equation is useful in the production of nuclear bombs as well as power from nuclear sources. The Manhattan project was directly concerned with the development of a weapon, but also contributed greatly to the development of science.

We all feel that technology can be misused. This has always been the case since man learned to master more or less the forces of nature. The Greeks and Romans were able, using a primitive and purely empirical technology, to damage the Mediterranean landscape to such an extent that eventually a great deal of the original soil was eroded. The rapid decay of the golden age of the Greeks and the more gradual decline of the Roman empire were at least in part caused by the irreversible destruction of the natural resources of wood, soil and water. The destructive potential of modern technology is tremendous as compared to the primitive empirical technology of the classical and medieval ages, since this technology is based on genuine knowledge about the real world. The question arises of who is responsible for the use and misuse of technology: scientists? or if not scientists, then who else? Before I go into this matter in more detail I would like to make two firm statements:

Firstly, we have gotten used to discussing the regressive phenomena of technology so exclusively that we have nearly forgotten how much we owe to technology. Many fancy statements about the hazards of technology show that the authors do not have the slightest idea of the circumstances under which our ancestors lived, how they suffered and how they died. The participants in the wild discussion about technology must learn to observe two elements which are essential for every fruitful discussion: fairness and competence.

Secondly, the favorite daydream of the bewildered citizen of stopping science, of calling a moratorium on science, is quite unrealistic; but even supposing it could be accomplished, such a solution of the problem would necessarily and rapidly lead to an irreversible decay of human culture. This statement is justified because the inevitable regressive phenomena of our technological culture can only be overcome by new technology. There is no way back. We have changed the world too much already. We have about 3.2 billion people too many on this planet to follow the romantic dream (even if it were attractive) of returning to the simple life close to nature. To repeat: As the situation is, there is no way back. The regressive phenomena of present-day technology, including pollution, can only be solved by progressive technology, and not by anything else. To call a moratorium on science would be to commit suicide, since the genuine knowledge which we need for the refined and improved technology of tomorrow, especially for a humanistic biological technology, is not yet available. It can only be elaborated by the efforts of science. On this basis alone any idea of a literal standstill in science is wrong-headed and cannot be taken seriously. Moreover, no scientist, and I hope no one who cares for the growth of knowledge, would accept a moratorium, even if it were practicable throughout a politically split world. The tradition of the free mind, that is, free inquiry and free publication, has been essential in setting the standards of truth in science; it has already been eroded by secrecy in government and industry, and we need to resist any extension of this [27].

The next step in my talk will bring us closer to the critical question. My thesis is that while all scientists agree that genuine knowledge is good in an ethical sense, the situation with respect to technology is totally different. Every technological achievement is necessarily ambivalent: that is, it can be good or bad, depending on one's point of view or on the particular situation. Technology is necessarily a double-edged tool. The classification of a given technological act or achievement as good or bad, right or wrong, is never certain because man is not omniscient. A given act can probably always be classified as ethically good or bad depending on the ends one has in mind, and depending on past, present and future conditions. The nuclear bombing of Hiroshima, which was intended to be and conceived of as an ethical act (to save the lives, both American and Japanese, which would be lost in a full-scale invasion), was later on classified as unethical [286].

To put this in more abstract terms, propositional functions of the form "X is good" have no verifiable meaning until they are modified to the form "I predict that X will be found good for the purpose of attaining end Y under conditions Z". The end Y and the conditions Z must be inferred and supplied before the statement makes sense, that is, before it acquires verifiable meaning. The situation as it is can be summarized as follows:

1. Every true proposition, every piece of genuine knowledge, can potentially be applied in technology.

2. Every technological system or measure is always and necessarily ambivalent.

3. It is a grievous misunderstanding of science to interpret the inherent moral and factual ambivalence of technology as an ambivalence of genuine knowledge. And it is bitterly ironical that this belief is widespread among the younger generation. This is a point I just do not understand since every real example you analyze will lead to the same conclusion: that the same piece of scientific knowledge can technologically be used to cure or to destroy, to erect a masterpiece of human culture or to destroy the beauty of nature. Modern medicine, or to be more specific, the successful application of the genuine knowledge of microbiology and immunobiology in biological technology, has decreased mortality at childbirth and during early childhood to nearly zero. This fact, which has been rightly praised as being one of the great achievements of modern medicine, has been by far the biggest contributor to the population explosion which will inevitably destroy our civilization if not stopped very soon.

The discovery of nuclear fission by HAHN and STRASSMANN in 1938 led not only to the atom bomb but also to the peaceful use of atomic energy of which many nations, for instance my own country, are in desperate need to increase the supply of electrical power and to decrease pollution. The atom bomb itself is a terrible weapon but the fears it causes may protect us against another big conventional war. I have barely survived the last conventional war in Europe, and I do not hesitate to praise atom bombs if their potential for destruction prevents another war of this type and these dimensions. If ethics are not powerful enough to prevent war (and they never have been in the past) we cannot but prevent wars by creating and maintaining strong fears. This is a bad solution, I agree, but it is much better than any type of conventional war.

In summarizing the statements I have made so far I shall return to the heart of the problem: The goal of science is knowledge, genuine knowledge, about the systems of the real world, including man. Absolute intellectual freedom is an indispensable prerequisite of science, and the scientific community must be allowed to pursue the search for truth according to their own ethical rules, the principles of which were established during the Renaissance and which have been obeyed since by every generation of scientists. The scientific community is responsible for the truth of scientific knowledge.

The goal of technology is to change the world according to decisions made by man. The data and laws of science are the most important elements in modern technology. Our present world is totally adapted to and dependent on technology, and thus technology is an irreversible phenomenon. The regressive phenomena of technology which threaten human life (and the life of every other creature as well) can only be overcome by the use and considered application of new technology which is again based on genuine scientific knowledge. Since many of these data are not yet available or not reliable enough to allow the formulation of laws, the progress of science is essential for our technological survival. Present-day scientific knowledge is clearly not satisfactory as a basis for future technology. This is especially true for the biological sciences and for the science of man.

Therefore, as I pointed out earlier, it would be a disaster for our own and every future generation if the rate of scientific progress were reduced. On the other hand, however, it is obvious that the technological application of scientific knowledge must be brought under rigorous control. Otherwise we shall destroy the surface of our little globe irreversibly, and this in the foreseeable future.

What institution could exert this control? What is the function of science and what is the responsibility of the individual scientist with regard to this critical problem? Science can advise man how to attain a goal but science cannot tell man whether he ought to choose that goal. In a free society which is necessarily pluralistic you will always find a multiplicity of goals, and for this reason any political decision is necessarily a compromise and can never please everybody. Some people feel that there is one goal which is common to all people, survival of man as an individual and as a species. Although many arguments have been advanced in favor of this position I doubt whether it can be derived from scientific knowledge. In any case, however, the assumption seems to be justified that the desire to survive as an individual and as a species is common to most people. If mankind were to accept this goal (which would require a political decision of gigantic dimensions), man would have to develop rapidly an appropriate moral system, that is a pattern of individual and collective conduct which would maximize the probability of survival of man as individual and species in a world which is ruled by technology. I feel that the explicit formulation of the new ethics which must be obeyed to attain this goal would be mainly up to the scientific community, because the genuine knowledge of science must replace traditional ideologies and established ethical propositions which are clearly outdated by the progress of our technological culture. Just as one example, people must be told by science that it is a sin from the point of view of survival to have more than two children and that the only ethical behavior in this connection is rigorous birth-control.

But let us return to reality: Does mankind really exist? Is mankind an institution we can trust to the same extent that we can trust the scientific community? Have the United Nations ever been able to solve a problem? Clearly, the problem of controlling technology is an international, world-wide task; it is a problem of mankind, probably the first problem common to all people. But mankind is still a fiction; in political reality, mankind does not exist. And there is not the slightest chance that this situation can be improved in the foreseeable future, even under the pressure exerted by the threat of world-wide pollution. For this reason, we must design control systems for technology which work on the level of existing political units, that is, on the level of a state or a nation, hoping that these models can eventually be extrapolated to larger and politically more complex systems.

Before going on to describe such a model I would like to confess frankly that I do not believe that man can survive for more than a brief period. The main reasons for my pessimism are the following:

1. Man is obviously not able to overcome his prescientific attitude of autistic thinking, as documented by the Bible, by which man made himself the center of the universe, created by God for his pleasure and temptation [248]. In accordance with this prescientific

attitude man became used to overpopulating, exploiting and polluting the earth and to destroying step by step and irreversibly the genetic and the system potential of evolution. I am afraid that we shall not be able to modify our anthropocentric attitude within the short period of time which is all that is available. Man has missed the right opportunity to overcome his prescientific approach to nature. Man should have changed his attitude towards nature when the new science-based technology made him the master of natural forces.

2. We observe an increase in irrationalism, illusion and day-dreaming throughout the western world just at the point when rationality is the main requirement for solving our problems. Instead of accepting the challenge and trying to master the problems of the modern world, a considerable part of the younger generation (at least in western Europe) tends to ignore reality, including the reality of human nature, and to indulge in dialectics and romanticism, sometimes combined with destructive criticism (which has always been the easy way). In any case, the art of clear thinking which is so characteristic of the scientific approach is being eroded rapidly as our universities incline more and more to a so-called "critical theory" which is hardly more than a vague anti-scientific ideology favoring MARX and HEGEL. The trick of this movement is to ignore scientific knowledge about nature as well as about man, to use nebulous terms and dialectic phrases and thus to escape from factual and logical criticism by the established sciences. What gave birth to science was a new way of thinking discovered in the Renaissance, the essence of which was the search for truth for truth's sake. But now we are about to sacrifice the strength of scientific thinking in favor of a new theology just at the point in history when genuine, unbiased knowledge and the full intellectual potential of the contemporary generation is required more than ever to secure the future of mankind.

After these admittedly personal remarks I want to return to my topic and to discuss the role of the scientist in making decisions as far as the development and application of technology are concerned.

Traditionally, decisions are made at several levels by politicians (involving the legislature, military men and administration) in accordance with the constitution of the particular country. Nowhere, to my knowledge, is the role of the scientist defined by the political and economic system and correspondingly his responsibility is not clear. Politicians usually emphasize, rightly, that political decisions are based on experience, too, namely on the accumulated experience of a long cultural tradition and evolution. However, this experience is not free from logical or factual contradictions nor is it organized into a fully rational pattern. It involves emotional and instinctive elements as well as mystical articles of faith which have become a part of ideologies and even of constitutions. Moreover, it will include conscious and subconscious precepts for wielding power, satisfying ambition and practicing demagogy. It seems that the extent of demagogy and irrationalism is always very great on the part of those who attack an established system. While this is natural, that is, dictated by the nature of man, many scientists (including myself) have become very allergic to this sort of manipulation in the course of time! Just one local innocuous example: In today's article in the *Daily Collegian* you can find the following statement: "Our revolution involves a destruction of power, not a change of those sitting in the throne of power". Does the author of this statement know that the human race has been trying in vain over thousands of years to achieve this goal? And does he know the theory developed by competent scientists to explain this failure? I hope he does not, and that he just expects that his potential readers do not know it either.

Now back to politics. Politicians are increasingly blamed by scientists and by technocrats for introducing these irrational elements into the course of decision-making. I feel that this criticism is not fully justified. Politics must necessarily involve irrational elements as long as we live in a free society and enjoy political pluralism, that is, have the right to choose our goal. The decision in favor of a particular goal as a rule involves a considerable degree of irrationalism, on the side of the citizen as well as on the side of the politician. I feel that this element should not be lost! The postulate that political freedom and pluralism should be sacrificed in order to make technocracy possible, can only be advanced by a person who does not know what dictatorship really means. A perfect dictatorship, that is, where all goals are set by one man or a few people, is a necessary prerequisite for a perfect technocracy. Therefore I plead against technocracy and in favor of political freedom and pluralism, even if we have to make allowances for irrationalities in political decisions at all levels. However, irrationalism has become extremely dangerous in a world which is equipped with a tremendous and even deadly technological power. Obviously, political irrationalism must be tamed, even within the constitutional framework of a free society. How can this be achieved?

I have proposed on several occasions [164, 167] the following model, which could be advanced in terms of general system theory but can also be explained quite simply. The model has been developed for the situation in my own country but it might be applicable, at least in principle, to every society which is based on personal freedom and on scientific technology. I firmly believe that the problems we are facing can only be solved by a rigorously organized cooperation between scientists (including scientifically trained technologists) and decision-makers (that is, political representatives of the people), wherein the responsibilities are clearly defined. In this model the decisions (including those about top research priorities) are to be made by the politicians but they may only be made between alternative models which are elaborated, or at least approved, by competent scientists. The scientist's responsibility is to ensure that only genuine knowledge is considered and the ethical code of science obeyed during the course of constructing the alternative models; this ensures, practically speaking, that every element and every relation in the model is reliable. On the other hand, the politician is responsible for the decision which is to be made between the different models. And at this point, irrationality (as pointed out previously) rightly comes into play.

Let me briefly describe the practical procedure of the two-step model of decision-making. Using different assumptions different self-consistent systems can be constructed, each of which is equally logical and equally justified by scientific knowledge. The alternative systems (or models) constructed by the scientists (including scientifically-minded technologists) are only different because different assumptions have been chosen. In most cases the solution of a complicated problem in the real world can be based on different assumptions (depending on the goals one has in mind), and a decision to use one assumption rather than another is required. In most cases the scientist cannot make the decision because several sets of assumptions are equally justified from the point of view of science. In those cases, a decision based on political experience, political taste, political prejudice must necessarily come into play. An example [210]: PTOLEMY placed the earth at the center of the universe, COPERNICUS the sun. Each system is perfectly logical and self-contained. COPERNICUS never ventured to give preference to his own system. The only way to decide between the two is to see where they lead if applied to a particular problem in the real world. We know now that an astronaut basing his calculations on PTOLEMY would disappear into nowhere. However, at the time of COPERNICUS nobody could have made a fully rational decision between PTOLEMY's and COPERNICUS' system if he had had the responsibility for a space program.

I would like to close my talk by discussing a few of the critical points of the cooperative model of decision-making.

1. A very serious problem in this cooperative model is the manner of communication between scientists and decision-makers. There are hardly any scientists among the political representatives, so they do not know the manner in which problems are solved in science. Our standard liberal arts education does not teach them. And neither do the law or business schools. On the other hand most scientists have no real appreciation of the difficulties of politics, they are, in general at least, politically naive, they do not understand how to exert political leadership, how to handle political power, to take over political responsibility and to live with the necessity of continuous moral and factual compromises. My feeling is that the rigorous application of system analysis and general system theory might be the only means to a fruitful cooperation. The detailed knowledge of the elements of the system is up to the specialized scientist or technologist, whereas the analysis of the relations between the elements is predominantly a logical problem which can be followed and understood by any man with a high I.Q. if he is willing to learn the language of system theory and if he is willing to admit that nowadays political problems must be handled like logical systems and that any intuitive solution of a problem must be backed by logical solutions which take into account all assumptions, the rational as well as the irrational ones. Let me emphasize again that I do not advocate banning irrationality from politics but I do advocate the necessity of defining clearly where, that is, at which points of the system, irrational elements come into play.

It is not very fair to blame the politicians for not being adjusted to the present situation. The scientific community in its ivory tower has not done much so far to insist on constant adjustment or to suggest means of improving this adjustment. The fact is that science and technology have rapidly created a new world without allowing time for adjustment of the educational and political systems. This ever-increasing discrepancy is a latent but very serious danger for the perpetuation of democracy and political order.

The understanding of scientific propositions by the non-scientist is further complicated by the fact that each noun in science has a double reference [286]. When the word "cell" is used, sometimes it refers to the conceptual model of the cell, other times it refers to the "thing-out-there", the postulated real system which the model attempts to represent. Without question, the semantic carelessness of scientists and the general lack of experience in the field of philosophy of science among scientists has considerably increased the difficulties of mutual understanding between scientists and non-scientists. This statement leads me directly to the second question:

2. Is a scientist able to cooperate with a politician at all? From my own experience I would say that by nature this cooperation is difficult; however, I firmly believe that it is possible if the duties and the responsibilities are clearly separated (as indicated in my model) and if both sides are really willing to cooperate towards a common goal. One great problem has been the fact that many distinguished scientists (and even some less distinguished ones) are sometimes ambitious and conceited beyond the average, and thus they often believe that a high degree of professional competence and excellence in a particular field of science will automatically lead to a general competence, even in the field of politics and political morals. However, experience tells us that the political performances of scientists

are amazingly poor, at least in general. This is understandable: the scientist's structure of thinking and arguing is so different from the politician's structure of thinking and acting that the scientists (including Nobel prize-winners) must perform poorly in the political arena if confronted with an experienced and intelligent political professional who knows how to employ his manipulative skills. If an established scientist tries to turn into a political professional, e. g. as a member of a legislature or as a Minister of Science and Technology, he will rapidly lose his scientific competence and stature and as a rule he will fall (after a surprisingly short time) between two stools: neither the professional politicians nor his fellow-scientists take him seriously any more. Therefore I plead for a cooperation where both the scientists and the politicians remain professionals in their particular fields and do not try to ignore or alternatively to take over the specific duties and responsibilities of the other partner.

3. Can the politician trust the scientist? Can he trust his models, his intellectual fairness; can he trust his political neutrality? This is, in my experience, the most difficult problem, and the politician has the right and even the duty to be sceptical. Science has developed excellent methods for deciding whether a given proposition is true or false, and every experienced scientist knows that the intellectual and experimental methods in many scientific fields are amazingly trustworthy. These methods undoubtedly produce genuine knowledge, if handled rightly. Rightly in this case means adhering to established scientific behavior, as given in the behavior code for scientists. This behavior code of science comprises a number of rough rules which must be obeyed in order to guarantee genuine knowledge: be honest, use only logic, never manipulate data, be fair, be without bias, do not make compromises, accept no authority except experience – these and a number of other rules are obeyed as a matter of course by every scientist whenever he works as a scientist, i.e. whenever he is concerned with obtaining or processing data. It has always amazed me to watch scientists (including myself) carefully and to notice how fast even those people who tend to be extremely biased, unreliable and even untruthful in their private lives will change their attitude as soon as they approach the workbench or the desk to do scientific work. Complete honesty is of course imperative in scientific work. Without discussing the ethical code of science any further I would just like to say that in the long run it pays the scientist to be absolutely honest, not only by not making false statements but by giving full expression to facts that are opposed to his views. Moral slovenliness is visited with far severer penalties in the scientific than in the business or political worlds [11].

In order to avoid any misunderstanding I would like to add the following restrictive remarks: scientists as a rule are not exceptionally ethical or modest in their personal lives, and there is no reason to expect them to behave better than or even differently from other people. Some scientists will even behave badly towards their fellow-scientists as soon as competition, priority, prestige and sometimes even money come into play. While these deficiencies are not liked very much they will eventually be ignored or tolerated by the scientific community, if the scientist in question does excellent work and has never violated his real reputation, which is for being absolutely honest and trustworthy in scientific matters, that is, in obtaining and handling data and in using logic. To summarize: a scientist does not have to fit into the same categories in his work and in his life, but if he does not, he must be content to exist as a dual personality. And he must be aware of this fact.

Against this background we again ask the question of whether the politician can really trust the scientific community to be both absolutely competent and absolutely honest in *scientific* matters. I have the unhappy feeling that at present the answer must be "no".

This is a very serious verdict and I need a few minutes to justify my statement. The problem involves at least two aspects: complexity and ideology.

Let me deal first with the aspect of complexity. We all realize, I guess, that certain fields of science which are concerned with very complex systems are still in their infancy. In these fields the formulation of general propositions, laws, is still a risk, if in fact it is possible at all. Even some important fields of biology must be placed in this category, including environmental sciences, psychology and the science of behavior. Let me briefly describe the principal situation in these fields, using the science of economics, a field of tremendous complexity, as an example [286]. The science of economics appears to be at a stage of development analogous to that of physics at the time of GALILEO, and it seems premature to try to construct a general economic model at this stage. In those communistic countries where the economy is based on general models derived in essence from a primitive prescientific ideology, the results so far have been very poor and sometimes even disastrous for the people. However, it is possible to isolate certain aspects of economics and to attempt to construct rigorous models of these isolated systems, models which can be checked scientifically, that is, by prediction and experience. This operations research has been regarded for some time as a very positive example of cooperation between science and government. While the reputation of some operations research groups has been marred by their predominant involvement in war planning, operations research as a whole has remained an outstanding example of fruitful cooperation between scientific experts and non-scientific decision-makers, even in one of the most complex and most complicated fields of science.

Let me comment briefly on the second aspect, the impact of science and ideology on our present European universities. A number of fields at our universities have been using the term science without deserving it, and as a consequence the formerly sharp distinction between speculative and scientific statements has rapidly disappeared. The high rank and reputation of the classical scientific fields is misused by these disciplines to support their own non- or semi-scientific statements. This is especially true for some fashionable disciplines within the social and political sciences but it is even true for parts of psychology and education. Many models developed in these fields contain more ideology than genuine knowledge about the nature of man. Therefore, the models do not work if accepted by the politician and applied to the real world. As an unhappy example from my own country I could mention some current models in the field of education, including the structure of our newly-founded universities. These repeated failures have diminished confidence in the reliability of science as a whole, among the public as well as among those politicians who are aware of their duties and responsibilities. It will be very hard to repair the damage already done to the reputation of science, just at a time when the full power of the true sciences is required more than ever to develop reliable models which could be used to overcome the terrific problems of our age. In any event, at present the bewildered citizen as well as the politician no longer has any reason to place a blind trust in scientific models. Rather he must take into account that they might be biased by ideology and even personal prejudice.

I shall close my lecture by summarizing the main points:

1. There are certain minimum requirements which a scientist must fulfill in order to be able to advise the politician within the framework of any cooperative model. The scientist must have a solid knowledge about the solid facts of a solid discipline. He must have sufficient knowledge in system analysis or general system theory. Also, he must have an innate drive to improve the social system. All these requirements must be met. A strong

motivation by itself is not sufficient. A scientist who does meet these requirements must therefore prevent the participation of others who have a strong desire to improve the world in the interests of science without having the capability of analyzing logically complicated situations and without being knowledgeable and experienced scientists.

2. The responsibilities of any member of the scientific community can be summarized as follows: We must preserve and rapidly increase the amount of genuine knowledge. We must re-establish the full trustworthiness of scientific propositions by eliminating completely the participation of those people and fields which do not rigorously obey the basic rules of the ethical code of science. We must contribute deliberately to the rationality of political decisions by thinking in terms of general system theory and by advising the politician, in accordance with the cooperative model, even at the expense of our scientific career in a particular field of research. Science is man's only means of survival. If we allow this marvelous instrument to be damaged or if we do not apply it rightly, man will disappear from this planet.

Suggested Further Reading

BEVERIDGE, W.I.B.: The Art of Scientific Investigation. New York: Vintage Books, Random House, 1950.
BRONOWSKI, J.: Science and Human Values. London: Hutchinson, 1961.
KUHN, T.S.: The Structure of Scientific Revolutions (2nd edit.) Chicago: The University of Chicago Press, 1970.
MOHR, H.: Wissenschaft und menschliche Existenz (2nd edit.) Freiburg: Rombach, 1970.
WALKER, M.: The Nature of Scientific Thought. Englewood Cliffs: Prentice-Hall, 1963.
ZIMAN, J.M.: Public Knowledge. London: Cambridge University Press, 1968.

Acknowledgements

The invaluable help of Mrs. SIAN FRICK who has checked the manuscript with respect to grammar and style is appreciated. I am further indebted to Mrs. UTE MEURER and Mrs. DORIS STACH for drawing of the figures and to Mrs. VIKTORIA KNITTEL-BERNHARDT and Mrs. SIAN FRICK for careful typing of the manuscript. I am grateful to Dr. P. SCHOPFER for critically reading the manuscript.

Literature Cited

1. AKOYUNOGLOU, G.A., SIEGELMAN, H.W.: Plant Physiol. *43*, 66 (1968).
2. BARR, T. C.: In: Evolutionary Biology, Vol. 2. Amsterdam: North Holland Publishing Company 1968.
3. BEEVERS, H.: Respiratory Metabolism in Plants. New York: Harper and Row 1961.
4. BERG, R.L.: Amer. Naturalist *43*, 103 (1959).
5. BERGFELD, R.: Z. Naturforsch. *18b*, 328 (1963).
6. BERGFELD, R.: Z. Naturforsch. *18b*, 558 (1963).
7. BERGFELD, R.: Z. Naturforsch. *22b*, 972 (1967).
8. BERNFIELD, M.R., WESSELS, N.K.: Develop. Biol. Supplement *4*, 195 (1970).
9. BERGMAN, K., BURKE, P.V., CERDA-OLMEDO, E., DAVID, C.N., DELBRÜCK, M., FORSTER, K.W., GOODELL, E.W., HEISENBERG, M., MEISSNER, G., ZALOKAR, M., DENNISON, D.S., SHROPSHIRE, W.: Phycomyces. Bacteriol. Rev. *33*, 99 (1969).
10. BERTSCH, W., MOHR, H.: Planta *65*, 17 (1965).
11. BEVERIDGE, W.I.B.: The Art of Scientific Investigation. New York: Vintage Books, Random House 1950.
12. BIENGER, I.: Dissertation, University of Freiburg, 1970.
13. BIENGER, I., SCHOPFER, P.: Planta *93*, 152 (1970).
14. BINDL, E., LANG, W., RAU, W.: Planta *94*, 156 (1970).
15. BIRTH, G.S.: Agr. Eng. *41*, 432 (1960).
16. BLONDON, F., JACQUES, R.: C.R. Acad. Sc. Paris *270*, 947 (1970).
17. BOARDMAN, N.K., THORNE, S.W., TREFFRY, T.E., ANDERSON, J.M., ROUGHAM, P.G.: In: Book of Abstracts, 2nd International Congress on Photosynthesis Research. (Stresa, Italy), p. 94 (1971).
18. BOISARD, J.: Physiol. Vég., *7*, 119 (1969).
18a. BOISARD, J., MARMÉ, D., SCHÄFER, E.: Planta *99*, 302 (1971).
19. BOISARD, J., SPRUIT, C.J.P., ROLLIN, P.: Meded. Landbouwhogeschool Wageningen 68-17 (1968).
20. BOLDT, A.: Master's thesis, University of Freiburg (1964).
21. BORTHWICK, H.A., HENDRICKS, S.B., TOOLE, E.H., TOOLE, V.K.: Bot. Gaz. *115*, 205 (1954).
22. BOYSEN-JENSEN, P.: Die Elemente der Pflanzenphysiologie. Jena: G. Fischer 1939.
23. BRADBEER, J.W.: In: Book of Abstracts, 2nd International Congress on Photosynthesis Research. (Stresa, Italy), p. 126 (1971).
24. BRANSCOMB, E.W., STUART, R.N.: Biochem. Biophys. Res. Commun. *32*, 731 (1968).
25. BRIGGS, W.R., CHON, H.P.: Plant Physiol. *41*, 1159 (1966).
26. BRITTEN, R.J., DAVIDSON, E.H.: Science *165*, 349 (1969).
27. BRONOWSKI, J.: In: The New York Times, October 18, 1971, p. 37[c].
28. BROOK, P.J.: Nature (Lond.) *222*, 390 (1969).
29. BÜNNING, E.: The Physiological Clock. New York: Academic Press 1963.
30. BURNS, R.G., INGLE, J.: Plant Physiol. *46*, 423 (1970).
31. BURT, C.: Brit. J. Psychol. *57*, 137 (1966).
32. BUTLER, W.L., HENDRICKS, S.B., SIEGELMAN, H.W.: Phytochem. Photobiol. *3*, 521 (1964).
33. BUTLER, W.L., HENDRICKS, S.B., SIEGELMAN, H.W.: Purification and properties of phytochrome. In: Chemistry and Biochemistry of Plant Pigments (T. W. GOODWIN, edit.). London: Academic Press 1965.
34. BUTLER, W.L., NORRIS, K.H., SIEGELMAN, H.W., HENDRICKS, S.B.: Proc. nat. Acad. Sci. *45*, 1703 (1959).
35. CALPOUZOS, L., CHANG, H.: Plant Physiol. *47*, 729 (1971).
36. CERFF, R.: Dissertation, University of Freiburg, 1971.
37. CHEN, D., OSBORNE, D.J.: Nature (Lond.) *225*, 336 (1970).

38. CLELAND, R.: Planta *98*, 1 (1971).
39. CORRELL, D.L., STEERS, E., TOWE, K.M., SHROPSHIRE, W.: Biochim. biophys. Acta *168*, 46 (1968).
40. CRACKER, L.E., STANDLEY, L.A., STARBUCK, M.J.: Plant Physiol. *48*, 349 (1971).
41. CUMMING, B.G.: Can. J. Bot. *41*, 901 (1963).
42. CUMMING, B.G., HENDRICKS, S.B., BORTHWICK, H.A.: Canad. J. Bot. *43*, 825 (1965).
43. CUMMING, B.G., WAGNER, E.: Ann. Rev. Plant Physiol. *19*, 381 (1968).
44. CURRY, G. M.: In: Physiology of Plant Growth and Development. (M. B. WILKINS, edit.), London: McGraw-Hill 1969.
45. v. DEIMLING, A., MOHR, H.: Planta *76*, 269 (1967).
46. DELBRÜCK, M., SHROPSHIRE, W.: Plant Physiol. *35*, 194 (1960).
47. DITTES, H.: personal communication (1971).
48. DITTES, H., MOHR, H.: Z. Naturforsch. *25b*, 708 (1970).
49. DITTES, L., RISSLAND, I., MOHR, H.: Z. Naturforsch. *26b*, 1175 (1971).
50. DOWNS, R., BORTHWICK, H.A.: Bot. Gaz. *117*, 310 (1956).
51. DOWNS, R.J., SIEGELMAN, H.W.: Plant Physiol. *38*, 25 (1963).
52. DRUMM, H., BRÜNING, K., MOHR, H.: Planta *106*, 259 (1972).
53. DRUMM, H., ELCHINGER, I., MÖLLER, J., PETER, K., MOHR, H.: Planta *99*, 265 (1971).
54. DRUMM, H., FALK, H., MÖLLER, J., MOHR, H.: Cytobiologie *2*, 335 (1970).
55. DRUMM, H., MOHR, H.: Planta *72*, 232 (1967).
56. DRUMM, H., MOHR, H.: Planta *75*, 343 (1967).
57. ELLIS, R.J., HARTLEY, M.R.: Nature (Lond.) New Biol. *233*, 193 (1971).
58. ENGELSMA, G.: Acta Bot. Neerl. *15*, 394 (1966).
59. ENGELSMA, G., VAN BRUGGEN, J.M.H.: Plant Physiol. *48*, 94 (1971).
60. EPSTEIN, H.T.: Elementary Biophysics. Reading, Mass.: Addison-Wesley 1963.
61. ETZOLD, H.: Planta *64*, 254 (1965).
62. EVANS, L.T. (edit.): Environmental Control of Plant Growth. New York: Academic Press 1963.
63. EVANS, L.T., KING, R.W.: Z. Pflanzenphysiol. *60*, 277 (1969).
64. EVERETT, M.S., BRIGGS, W.R.: Plant Physiol. *45*, 679 (1970).
65. FALK, H., STEINER, A.M.: Naturwiss. *55*, 500 (1968).
66. FEIERABEND, J.: Planta *71*, 326 (1966).
67. FILNER, P., VARNER, J.E.: Proc. nat. Acad. Sci. *58*, 1520 (1967).
68. FONDEVILLE, J.C., BORTHWICK, H.A., HENDRICKS, S.B.: Planta *69*, 359 (1966).
69. FRANKLAND, B., HARTMANN, K.M.: personal communication (1969).
70. FRÉDÉRICQ, H.: Plant Physiol. *39*, 182 (1964).
71. FRIEDERICH, K.E.: Planta *84*, 81 (1969).
72. FURUYA, M.: Biochemistry and physiology of phytochrome. In: Progress in Phytochemistry, Vol. 1. London: Interscience Publishers 1968.
73. GALSTON, A. W.: In: Handbuch der Pflanzenphysiologie (W. H. RUHLAND, edit.), Vol. 17/1, p. 492. Berlin-Göttingen-Heidelberg: Springer 1959.
74. GALSTON, A.W.: Proc. nat. Acad. Sci. *61*, 454 (1968).
75. GALSTON, A.W., DAVIES, P.J.: Science *163*, 1288 (1969).
76. GALSTON, A.W., DAVIES, P.J.: Control Mechanisms in Plant Development. Englewood Cliffs: Prentice-Hall 1970.
77. GARDNER, G., PIKE, C.S., RICE, H.V., BRIGGS, W.R.: Plant Physiol. *48*, 686 (1971).
78. GARN, S.M.: Human Races. (2nd edit.), Springfield: Thomas 1962.
79. GAUDILLERE, J.P., COSTES, C.: Photosynthetica *5*, 272 (1971).
80. GEISER, N.: Master's thesis, University of Freiburg, 1964.
81. GOODWIN, T.W.: In: Biochemistry of Chloroplasts, Vol. 2. (T.W. GOODWIN, edit.), New York: Academic Press 1967.
82. GOODWIN, T. W.: In: Structure and Function of Chloroplasts. (M. GIBBS, edit.) Berlin-Heidelberg-New York: Springer 1971.
83. DE GREEF, J., BUTLER, W.L., ROTH, T.F.: Plant Physiol. *47*, 457 (1971).
84. DE GREEF, J., BUTLER, W.L., ROTH, T.F., FRÉDÉRICQ, H.: Plant Physiol. *48*, 407 (1971).
85. GRESSEL, J., STRAUSBAUCH, L., GALUN, E.: Nature (Lond.) *232*, 648 (1971).
86. HACHTEL, W.: Planta *102*, 247 (1972).

87. HÄCKER, M.: Planta 76, 309 (1967).
88. HÄCKER, M., HARTMANN, K.M., MOHR, H.: Planta 63, 253 (1964).
89. HAHLBROCK, K., WELLMANN, E.: Planta 94, 236 (1970).
90. HÄMMERLING, J.: Ann. Rev. Plant Physiol. 14, 65 (1963).
91. HAMNER, K.: In: Environmental Control of Plant Growth. (L.T. EVANS, edit). New York: Academic Press 1963.
92. HANKE, J., HARTMANN, K.M., MOHR, H.: Planta 86, 235 (1969).
93. HARTMANN, E.: Planta 101, 159 (1971).
94. HARTMANN, K.M.: Photochem. Photobiol. 5, 349 (1966).
95. HARTMANN, K.M.: Naturwiss. 54, 544 (1967).
96. HARTMANN, K.M.: Photoreceptor problems in photomorphogenic responses under high-energy-conditions (UV-blue-far-red). In: Book of Abstracts. European Photobiology Symposium, Hvar (Jugoslavia), p. 29, (1967).
97. HARTMANN, K.M.: Z. Naturforsch. 22b, 1172 (1967).
98. HAUPT, W.: Z. Pflanzenphysiol. 58, 331 (1968).
99. HAUPT, W.: Physiol. vég. 8, 551 (1970).
100. HAUPT, W.: Z. Pflanzenphysiol. 62, 287 (1970).
101. HAUPT, W., BOCK, G.: Planta 59, 38 (1962).
102. HAUPT, W., MÖRTEL, G., WINKELNKEMPER, I.: Planta 88, 183 (1969).
103. HENDRICKS, S.B.: Science 141, 21 (1963).
104. HENDRICKS, S.B.: Photochemical aspects of plant photoperiodicity. In: Photophysiology, Vol. 1 (A. C. GIESE, edit.). London: Academic Press 1964.
105. HENDRICKS, S.B., BORTHWICK, H.A.: The physiological functions of phytochrome. In: Chemistry and Biochemistry of Plant Pigments (T.W. GOODWIN, edit.). London: Academic Press 1965.
106. HENDRICKS, S.B., BUTLER, W.L., SIEGELMAN, H.W.: J. Phys. Chem. 66, 2550 (1962).
107. HENDRICKS, S.B., SIEGELMAN, H.W.: Phytochrome and photoperiodism in plants. In: Comparative Biochemistry, Vol. 25. (M. FLORKIN and E. H. STOTZ, edit.), Amsterdam: Elsevier 1967.
108. HILL, A.V.: Nature (Lond.) 113, 859 (1924).
109. HILLMAN, W.S.: The physiology of phytochrome. Ann. Rev. Plant Physiol. 18, 301 (1967).
110. HILLMAN, W.S.: In: Physiology of Plant Growth and Development (M.B. WILKINS, edit.). London: McGraw-Hill 1969.
111. HILLMAN, W.S., KOUKKARI, W.L.: Plant Physiol. 42, 1413 (1967).
112. HOCK, B., KÜHNERT, E., MOHR, H.: Planta 65, 129 (1965).
113. HOCK, B., MOHR, H.: Planta 65, 1 (1965).
114. IMASEKI, H., PJON, C.J., FURUYA, M.: Plant Physiol. 48, 241 (1971).
115. JACOBSON, J.V., SCANDALIOS, J.G., VARNER, J.E.: Plant Physiol. 45, 367 (1970).
116. JAFFE, M.J.: Science 162, 1016 (1968).
117. JAFFE, M.J.: Plant Physiol. 46, 768 (1970).
118. JAFFE, M.J.: personal communication (1971).
119. JAFFE, M.J., GALSTON, A.W.: Planta 77, 135 (1967).
120. JAKOBS, M., MOHR, H.: Planta 69, 187 (1966).
121. JANYSTIN, B., DRUMM, H.: Naturwiss. 59, 218 (1972).
122. JENSEN, A.R.: Harv. Educ. Rev. 39, 1 (1969).
123. JOLIOT, P., KOK, B.: Biochim. biophys. Acta 153, 635 (1968).
124. JOST, J., RICKENBERG, H.V.: Ann. Rev. Biochem. 40, 741 (1971).
125. KAHN, A., BOARDMAN, N.K., THORNE, S.W.: J. Mol. Biol. 48, 85 (1970).
126. KALOW, W.: Pharmacogenetics. Philadelphia: Saunders 1962.
127. KAROW, H., MOHR, H.: Planta 72, 170 (1967).
128. KASEMIR, H.: personal communication (1971).
129. KASEMIR, H.: personal communication (1972).
130. KASEMIR, H., MOHR, H.: Planta 67, 33 (1965).
131. KASEMIR, H., MOHR, H.: Plant Physiol. (1972, in press).
132. KENDRICK, R.E., FRANKLAND, B.: Planta 82, 317 (1968).
133. KENDRICK, R.E., FRANKLAND, B.: Planta 85, 326 (1969).
134. KENDRICK, R.E., FRANKLAND, B.: Planta 86, 21 (1969).
135. KENDRICK, R.E., SPRUIT, C.J.P., FRANKLAND, B.: Planta 88, 293 (1969).

136. KIRK, J.T.O., TILNEY-BASSETT, R.A.E.: The Plastids. London: Freeman 1967.
137. KLEIBER, H., MOHR, H.: Z. Bot. 52, 78 (1963).
138. KÖNITZ, W.: Planta 51, 1 (1958).
139. KOUKKARI, W.L., HILLMAN, W.S.: Plant Physiol. 43, 698 (1968).
140. KROES, H.H.: Meded. Landbouwhogeschool Wageningen 70-18 (1970).
141. KÜHN, A.: Grundriß der Vererbungslehre. Heidelberg: Quelle und Meyer 1961.
142. LAMPORT, D.T.A.: Ann. Rev. Plant Physiol. 21, 235 (1970).
143. LANGE, H., BIENGER, I., MOHR, H.: Planta 76, 359 (1967).
144. LANGE, H., MOHR, H.: Planta 67, 107 (1965).
145. LANGE, H., SHROPSHIRE, W., MOHR, H.: Plant Physiol. 47, 649 (1971).
146. LUKENS, R.J.: Phytopathology 55, 1032 (1965).
147. MALCOSTE, R.: C.R. Acad. Sc. Paris 267, 613 (1968).
148. MALCOSTE, R.: C.R. Acad. Sc. Paris 269, 701 (1969).
149. MARMÉ, D.: Planta 88, 43 (1969).
150. MARMÉ, D., MARCHAL, B., SCHÄFER, E.: Planta 100, 331 (1971).
151. MARMÉ, D., SCHÄFER, E., TRILLMICH, F., HERTEL, R.: In: Book of Abstracts. European Ann. Symposium on Plant Photomorphogenesis, Athens-Eretria (Greece) p. 36 (1971).
152. MARMÉ, D., SCHÄFER, E.: Z. Pflanzenphysiol. 67, 192 (1972).
153. MASONER, M., UNSER, G., MOHR, H.: Planta 105, 267 (1972).
154. MCARTHUR, J.A., BRIGGS, W.R.: Plant Physiol. 48, 46 (1971).
155. MCCLINTOCK, B.: Genetic systems regulating gene expression during development. Develop. Biol. Supplement 1, 84 (1967).
156. MENZEL, H.: personal communication (1967).
157. MOHR, H.: Planta 46, 534 (1956).
158. MOHR, H.: Planta 47, 127 (1956).
159. MOHR, H.: Planta 49, 389 (1957).
160. MOHR, H.: In: Environmental Control of Plant Growth. (L.T. EVANS, edit.), p. 262, New York: Academic Press 1963.
160a. MOHR, H.: J. Linn. Soc. (Bot.) 58, 287 (1963).
161. MOHR, H.: Ber. dtsch. bot. Ges. 78, 54 (1965).
162. MOHR, H.: Z. Pflanzenphysiol. 54, 63 (1966).
163. MOHR, H.: Photochem. Photobiol. 5, 469 (1966).
164. MOHR, H.: Freiburger Universitätsreden, Neue Folge, Heft 46 (1969).
165. MOHR, H.: Biologie als quantitative Wissenschaft. Beilage zu: Naturwiss. Rdsch., Heft 7, 779 (1970).
166. MOHR, H.: Naturw. Rdsch. 23, 187 (1970).
167. MOHR, H.: Grenzgebiete der Wissenschaft 19, 298 (1970).
168. MOHR, H., APPUHN, U.: Planta 59, 49 (1962).
169. MOHR, H., APPUHN, U.: Planta 60, 274 (1963).
170. MOHR, H., BIENGER, I.: Planta 75, 180 (1967).
171. MOHR, H., BIENGER, I., LANGE, H.: Nature 230, 56 (1971).
172. MOHR, H., HOLDERIED, CH., LINK, W., ROTH, K.: Planta 76, 348 (1967).
173. MOHR, H., OHLENROTH, K.: Planta 57, 656 (1962).
174. MOHR, H., PETERS, E.: Planta 55, 637 (1960).
175. MOHR, H., SCHLICKEWEI, I., LANGE, H.: Z. Naturforsch. 20b, 819 (1965).
176. MOHR, H., SENF, R.: Planta 71, 195 (1966).
176a. MOHR, H., SITTE, P.: Molekulare Grundlagen der Entwicklung. München: BLV 1971.
177. MONTAGUE, M.F.A.: An Introduction to Physical Anthropology. Springfield: Thomas 1960.
178. MUDD, J.B., MCMANUS, T.T., ONGUN, A., MCCULLOGH, T.E.: Plant Physiol. 48, 335 (1971).
179. MUMFORD, F.E.: Biochemistry 5, 522 (1966).
180. MUMFORD, F.E., JENNER, E.L.: Biochemistry 5, 3657 (1966).
181. MUMFORD, F.E., JENNER, E.L.: Biochemistry 10, 98 (1971).
182. NABORS, M.W., LANG, A.: Planta 101, 1 (1971).
183. NABORS, M.W., LANG, A.: Planta 101, 26 (1971).
184. NADLER, K., GRANICK, S.: Plant Physiol. 46, 240 (1970).
185. NAKAYAMA, S.: Ecol. Rev. 14, 325 (1958).

186. NAKAYAMA, S., BORTHWICK, H.A., HENDRICKS, S.B.: Bot. Gaz. *121*, 237 (1960).
187. NEWMAN, I.A., BRIGGS, W.R.: Plant Physiol. *47* (supplement), p. 1/6 (1971).
188. NITSCH, J.P.: Proc. Amer. Soc. hort. Sci. *70*, 512 (1957).
189. OBERDORFER, U., KASEMIR, H.: personal communication (1971).
190. OELZE-KAROW, H.: personal communication (1971).
191. OELZE-KAROW, H., BUTLER, W.L.: Plant Physiol. *48*, 621 (1971).
192. OELZE-KAROW, H., MOHR, H.: Z. Naturforsch. *25b*, 1282 (1970).
193. OELZE-KAROW, H., SCHOPFER, P., MOHR, H.: Proc. nat. Acad. Sci. *65*, 51 (1970).
194. OLTMANNS, F.: Morphologie und Biologie der Algen. Jena: G. Fischer 1922.
195. ONGUN, A., MUDD, J.B.: J. biol. Chem. *243*, 1558 (1968).
196. PARKER, M.W., HENDRICKS, S.B., BORTHWICK, H.A., SCULLY, N.J.: Botan. Gaz. *108*, 1 (1946).
197. PAYER, D.: Planta *86*, 103 (1969).
198. PAYER, H.D., MOHR, H.: Planta *86*, 286 (1969).
199. PAYER, H.D., SOTRIFFER, U., MOHR, H.: Planta *85*, 270 (1969).
200. PFEFFER, W.: Pflanzenphysiologie. Leipzig: Wilhelm-Engelmann-Verlag 1904.
201. PJON, C.J., FURUYA, M.: Plant Cell Physiol. *8*, 709 (1967).
202. PORTER, G.: Photochemistry of complex molecules. In: An Introduction to Photobiology (C.P. SWANSON, edit.). Englewood Cliffs: Prentice-Hall 1969.
203. VAN POUCKE, M., BARTHE, F., MOHR, H.: Naturwiss. *56*, 417 (1969).
204. VAN POUCKE, M., BARTHE, F.: Planta *94*, 308 (1970).
205. VAN POUCKE, M., CERFF, R., BARTHE, F., MOHR, H.: Naturwiss. *56*, 132 (1970).
206. PRATT, L.H., BRIGGS, W.R.: Plant Physiol. *41*, 467 (1966).
207. PRATT, L.H., BUTLER, W.L.: Plant Physiol. Abstracts, No. *86* (1969).
208. PRATT, L.H., BUTLER, W.L.: Photochem. Photobiol. *11*, 361 (1970).
209. PRATT, L.H., COLEMAN, R.A.: Proc. nat. Acad. Sci. *68*, 2431 (1971).
210. QUAIL, P.: personal communication (1972).
211. RAGHAVAN, V.: Planta *81*, 38 (1968).
212. RAGHAVAN, V.: Physiol. Plantarum *21*, 1020 (1968).
213. RAGHAVAN, V., DEMAGGIO, E.A.: Plant Physiol. *48*, 82 (1971).
214. RAU, W.: Planta *72*, 14 (1967).
215. REBEIZ, C.A., CASTELFRANCO, P.A.: Plant Physiol. *47*, 33 (1971).
216. RISSLAND, I., MOHR, H.: Planta *77*, 239 (1967).
217. ROBINSON, G.A., BUTCHER, R.W., SUTHERLAND, E.W.: Ann. Rev. Biochem. *37*, 149 (1968).
218. ROLLIN, P.: Bull. Soc. Fr. Physiol. veg. *14*, 47 (1968).
219. ROLLIN, P.: Phytochrome, Photomorphogénèse et Photopériodisme. Paris: Masson 1970.
220. ROLLIN, P., MALCOSTE, R., EUDE, D.: Planta *91*, 227 (1970).
221. ROTH, K.: Master's thesis, University Freiburg, 1968.
222. ROTH, K., LINK, W., MOHR, H.: Cytobiologie *1*, 248 (1970).
223. RUBERY, P.H., FOSKET, D.E.: Planta *87*, 54 (1969).
224. RUBERY, P.H., NORTHCOTE, D.H.: Nature (Lond.) *219*, 1230 (1968).
225. RÜDIGER, W.: Liebigs Ann. Chem. *723*, 208 (1969).
226. RUSSELL, D.W.: Published Abstracts. XI. International Botanical Congress, Seattle (Wash.), p. p. 185 (1969).
227. SALISBURY, F.B.: The Flowering Process. Oxford: Pergamon Press 1963.
228. SALMAN, A.G.F.: Planta *101*, 117 (1971).
229. SATTER, R.L., GALSTON, A.W.: Science *174*, 518 (1971).
230. SATTER, R.L., GALSTON, A.W.: Plant Physiol. *48*, 470 (1971).
231. SATTER, R.L., MARINOFF, P., GALSTON, A.W.: Amer. J. Bot. *57*, 916 (1970).
232. SATTER, R.L., SABNIS, D.D., GALSTON, A.W.: Amer. J. Bot. *57*, 374 (1970).
233. SCARR-SALAPATEK, S.: Science *174*, 1223 (1971).
234. SCARR-SALAPATEK, S.: Science *174*, 1285 (1971).
235. SEITZ, U., RICHTER, G.: Planta *92*, 309 (1970).
236. SHROPSHIRE, W.: Physiol. Rev. *43*, 39 (1963).
237. SHROPSHIRE, W., MOHR, H.: Photochem. Photobiol. *12*, 145 (1970).
238. SIEGELMAN, H.W.: Phytochrome. In: Physiology of Plant Growth and Development. (M.B. WILKINS, edit.), London: McGraw-Hill 1969.

239. SISLER, E.C., KLEIN, W.H.: Physiol. Plantarum *16*, 315 (1963).
240. SMITH, H.: Nature (Lond.) *227*, 665 (1970).
241. SPRUIT, C.J.P., KENDRICK, R.E.: Planta *103*, 319 (1972).
242. SPRUIT, C.J.P., MANCINELLI, A.L.: Planta *88*, 303 (1969).
243. SPRUIT, C.J.P., RAVEN, C.W.: Acta Bot. Neerl. *19*, 165 (1970).
244. STEINER, A.M.: Naturwiss. *54*, 497 (1967).
245. STEINER, A.M.: Planta *82*, 223 (1968).
246. STEINER, A.M.: Z. Pflanzenphysiol. *59*, 401 (1968).
247. STERN, H.: In: Regulateurs naturels de la croissance végétale. Paris: Centre Nat. Rech. Sci., 1964.
248. SZENT-GYORGYI, A.: In: The New York Times, October 23, 1971.
249. SCHÄFER, E., MARCHAL, B., MARMÉ, D.: Planta *101*, 265 (1971).
250. SCHÄFER, E., MARCHAL, B., MARMÉ, D.: Photochem. Photobiol. *15*, 457 (1972).
251. SCHEIBE, J., LANG, A.: Plant Physiol. *40*, 485 (1965).
252. SCHIMKE, R.T.: In: Current Topics in Cellular Regulation, Vol. 1, 77 (1969).
253. SCHNEIDER, M.J., STIMSON, W.R.: Plant Physiol. *48*, 312 (1971).
254. SCHOPFER, P.: Planta *69*, 158 (1966).
255. SCHOPFER, P.: Planta *72*, 297 (1967).
256. SCHOPFER, P.: Planta *72*, 306 (1967).
257. SCHOPFER, P.: Planta *74*, 210 (1967).
258. SCHOPFER, P.: Planta *85*, 383 (1969).
259. SCHOPFER, P.: Experimente zur Pflanzenphysiologie. Freiburg: Rombach 1970.
260. SCHOPFER, P.: Planta *99*, 339 (1971).
261. SCHOPFER, P.: personal communication (1971).
262. SCHOPFER, P.: In: Nucleic Acids and Proteins in Higher Plants. Proc. Symp. Tihany (Hungary), (Sept. 1971), in press.
263. SCHOPFER, P., HOCK, B.: Planta *96*, 248 (1971).
264. SCHOPFER, P., MOHR, H.: Plant Physiol. *49*, 8 (1972).
265. SCHOPFER, P., OELZE-KAROW, H.: Planta *100*, 167 (1971).
266. SCHNARRENBERGER, C., MOHR, H.: Planta *75*, 114 (1967).
267. SCHNARRENBERGER, C., MOHR, H.: Planta *94*, 296 (1970).
268. TANADA, T.: Plant Physiol. *43*, 2070 (1968).
269. TAYLORSON, R.B., HENDRICKS, S.B.: Plant Physiol. *47*, 619 (1971).
270. TEZUKA, T., YAMAMOTO, Y.: Bot. Mag. Tokyo *82*, 130 (1969).
271. TOLBERT, N.E.: Ann. Rev. Plant Physiol. *22*, 45 (1971).
272. TOMKINS, G.M., GELEHRTER, Th. D., GRANNER, D., MARTIN, D., SAMUELS, H.H., THOMPSON, E.B.: Science *166*, 1474 (1969).
273. TOPPER, Y.J., FRIEDBERG, S.H., OKA, T.: Developmental Biol. Supplement *4*, 101 (1970).
274. TRAVIS, R.L., HUFFAKER, R.C., KEY, J.L.: Plant Physiol. *46*, 800 (1970).
274a. TRAVIS, R.L., KEY, J.L.: Plant Physiol. *48*, 617 (1971).
275. TRELEASE, R.N., BECKER, W.M., GRUBER, P.J., NEWCOMB, E.H.: Plant Physiol. *48*, 461 (1971).
276. TRÉMOLIÈRES, A.: Ann. Biol. *9*, 113 (1970).
277. UNSER, G., MASONER, M.: Naturwiss. *59*, 39 (1972).
278. UNSER, G., MOHR, H.: Naturwiss. *57*, 358 (1970).
279. VINCE, D.: Physiol. Plantarum *18*, 474 (1965).
280. VINCE, D.: J. Royal Horticult. Soc. *45*, 214 (1970).
281. WADDINGTON, C.H.: Science *166*, 639 (1969).
282. WAGNÉ, C.: Physiol. Plantarum *17*, 751 (1964).
283. WAGNER, E., BIENGER, I., MOHR, H.: Planta *75*, 1 (1967).
284. WAGNER, E., FROSCH, S.: personal communication (1971).
285. WAGNER, E., MOHR, H.: Planta *71*, 204 (1966).
286. WALKER, M.: The Nature of Scientific Thought. Englewood Cliffs: Prentice-Hall 1963.
287. WALTER, A., SHROPSHIRE, W.: Diss. Abstracts *29*, 499 (1969).
288. WEIDNER, M.: Planta *75*, 94 (1967).
289. WEIDNER, M., JAKOBS, M., MOHR, H.: Z. Naturforsch. *20b*, 689 (1965).
290. WEIDNER, M., MOHR, H.: Planta *75*, 99 (1967).
291. WEIDNER, M., MOHR, H.: Planta *75*, 109 (1967).

292. WEISS, P.: The Science of Biology. New York: McGraw-Hill 1967.
293. WELLBURN, F.A.M., WELLBURN, A.R.: Biochem. Biophys. Res. Comm. *45*, 747 (1971).
294. WELLMANN, E.: Planta *101*, 283 (1971).
295. WELLMANN, E.: personal communication (1971).
296. WILKINS, H.: Evolution *25*, 530 (1971).
297. WILSON, R.S.: Science *175*, 914 (1972).
298. WITHROW, R.B., KLEIN, W.H., ELSTAD, V.: Plant Physiol. *32*, 453 (1957).
299. WITT, H.T., RUMBERG, B., SCHMIDT-MENDE, P., SIGGEL, U., SKERRA, B., VATER, J., WEIKARD, J.: Angew. Chemie *77*, 821 (1965).
300. WOLFF, P.H.: Science *175*, 449 (1972).
301. YIN, H.C., FAN, I.J., SHEN, G.M., LI, T.Y., SHEN, Y.K.: Sci. sinica *14*, 1184 (1965).
302. ZAMENHOF, S., EICHHORN, H.H.: Nature (Lond.) *216*, 456 (1967).
303. ZUCKER, M.: Plant Physiol. *47*, 442 (1971).

Subject Index

Acetylcholine 114, 115
Actinomycin D. 56
Action spectrum 5
Albizzia julibrissin 84, 86
Amaranthus caudatus 21, 187, 188
– *retroflexus* 187
δ-amino-levulinic acid 172
AMP, cyclic 113
Amylase 118–121, 149
Anthocyanin synthesis 10, 37, 38, 45, 49, 55–57, 81, 107, 108, 125–129, 138, 139, 141, 144, 147, 148, 193
Arabidopsis thaliana 188
Ascorbate oxidase 38, 66, 110, 111, 129, 149, 152–154
Ascorbic acid accumulation 77–79, 82, 108–110, 130

Bacillus subtilis 155
Beltsville group 5, 13, 84, 179

Carbohydrate metabolism 95–96
Carotenoid accumulation 79–81, 161–166, 190
Catalase 67, 68, 175–177
Catalpa bignonioides 184
Cell growth 90–96
Chenopodium album 114
– *amaranticolor* 180
– *rubrum* 114, 181, 182
Chlamydomonas eugametos 7
Chloramphenicol 56, 147, 148, 166
Chlorophyll a 166–168
– – accumulation 46, 170–175
Chlorophyll bleaching 11
Chloroplast 160, 161, 199, 201
– movement 29–33, 83
Cinnamic acid hydroxylase 143, 193
Coleoptile 33, 40
Coleoptile, segments of 103
Cotyledons 159
–, enlargement of 57–58
Cyanidin 142
Cycloheximide 56, 166
Cytodifferentiation 141

Development 5
Developmental homeostasis 212

Dichroism 24, 33
Difference spectrum 13
Differentiation, primary 49, 51, 132–141, 144–146, 188
–, secondary 51, 132–141, 144–146
Digalactosyl diglyceride 166, 167
Diphenylamine 190
Dormancy 183, 187
Dryopteris filix-mas 25–27, 194–204

Effectors 141
Endogenous rhythms 179–183
Energetics 153–156
Enzyme induction 60–68
– repression 69–76
– synthesis 129–130
Environmental variance 214
Ethylene 104
Etiolation 155
Evolution 155

Fagopyrum esculentum 208
Fern gametophytes 194–204
Flavone glycosides 144
Flavonoid synthesis 141–148
Flavoprotein 190–192, 197, 206, 207
Flower initiation 178–184
Funaria hygrometrica 29
Fusarium aquaeductuum 190, 191

Galactolipids 166–168
Galactosyltransferase 168, 169
Gene activity, differential 54
Gene repression, differential 55
Genetic information, phenotypization of 213–215
Genetic variance 214
Gentiana campestris 211
Gibberellic acid 88–90, 118, 120
Glyceraldehyde – 3 – phosphate dehydrogenase 66, 169, 170, 181, 182
Glycine max 179, 180
Glycollate oxidase 64–66, 149, 175, 176
Glyoxylate reductase 175, 176
Glyoxysomes 175, 177
Growth 88, 141
Growth-limiting proteins 103

Plant Growth Substances 1970

Proceedings of the
7th International Conference
on Plant Growth Substances
Held in Canberra, Australia,
December 7 - 11, 1970

Edited by D.J. Carr,
Canberra City, Australia

461 fig. XIV, 837 pages. 1972

From the preface

*The aim of these meetings
is to provide a forum
for discussion of new work
and recent trends.
While it would be idle
to suppose that a
conference of this kind
could deal with all
the many aspects of research
on plant growth substances
going on all over the world,
readers will find in
this volume a coverage of
some of the more
exciting topics currently
under investigation.
Since the last conference
there has been a very
large increase in the known
number of naturally-
occurring plant-growth
substances, especially in the
classes of gibberellins
(now approaching 40),
inhibitors and cytokinins,
and it is inconceivable
that research on their effects
could expand to fit the rate
of discovery.
The interaction between
studies of growth hormones
and those of molecular
biology is growing
in intensity. It is perhaps
too early to expect tangible
results of this interaction
but tactics and initial
skirmishes in the struggle
to understand the
mechanisms of hormone
action are recorded in the
volume. Many natural-
products chemists have
embraced the study of plant
hormones and brought with
them expertise in automated
structural analysis,
and separation techniques.
A special session was
devoted to two papers on
applications of gasliquid
chromatography and
mass spectroscopy to plant
hormone research.*

Contents

**Springer-Verlag
Berlin
Heidelberg
New York**

London · München · Paris
Sydney · Tokyo · Wien

Microautoradiography and Electron Probe Analysis: Their Application to Plant Physiology

Editor: **U. Lüttge,**
Technische Hochschule
Darmstadt

With 78 figures
Approx. 280 pages. 1972

This book is a practical introduction to the methods of microautoradiography and electron probe analysis in experimentation, preparation, and analysis. Numerous examples are given.

Contents:

U. Lüttge, Botanical Applications of Micro-autoradiography. – P. Dörmer, Photometric Methods in Quantitative Autoradiography. – J. B. Passioura, Quantitative Autoradiography in the Presence of Crossfire. – U. Lüttge, Microautoradiography of Water-Soluble Inorganic Jons. – E. Eschrich, E. Fritz, Micro-autoradiography of Water-Soluble Organic Compounds. – R. G. Herrmann, W. O. Abel, Microautoradiography of Organic compounds Insoluble in a Wide Range of Polar and Non-polar Solvents. – J. Pickett-Heaps, Auto-radiography with the Electron Microscope: Experimental Techniques and Considerations Using Plant Tissues. – A. Läuchli, Electron Probe Analysis.

Springer-Verlag
Berlin · Heidelberg · New York
London München Paris Sydney Tokyo Wien